Unleashing the Power of UX Analytics

Proven techniques and strategies for uncovering user insights
to deliver a delightful user experience

Jeff Hendrickson

BIRMINGHAM—MUMBAI

Unleashing the Power of UX Analytics

Group Product Manager: Alok Dhuri
Publishing Product Manager: Uzma Sheerin
Book Project Manager: Deeksha Thakkar
Senior Editor: Rounak Kulkarni
Technical Editor: Jubit Pincy
Copy Editor: Safis Editing
Proofreader: Safis Editing
Indexer: Tejal Daruwale Soni
Production Designer: Jyoti Chauhan
Developer Relations Marketing Executives: Deepak Kumar and Mayank Singh
Business Development Executive: Puneet Kaur

First published: September 2023

Production reference: 1270723

Published by Packt Publishing Ltd.
Grosvenor House
11 St Paul's Square
Birmingham
B3 1RB

ISBN 978-1-80461-474-7

www.packtpub.com

To my parents, Dottie and Leroy Hendrickson, for their never-ending support throughout the years, and for always believing in me, no matter where my creative soul took me. Thanks, Mom and Dad, for the years of support and love.

To my sons, Miles and Dylan, two fine young men who are forging their own paths in the world – I'm so proud of you both and am eternally grateful to be your father.

And to all those I've worked with around the world over the years in building this career in UX. You were my champions, my coaches, and my collaborators, and I thank you for your support.

– Jeff Hendrickson

Foreword

We got it!

The month started off as usual, with me planning my workload and preparing for a special assignment with a customer and our new lead UX engineer. We had customer calls, product management calls, and we spent time organizing and planning. The new UX person was calm, cool, and collected, like he had done this many times before.

We landed on one of the hottest days in the southern, gulf coast part of Texas and went straight to the customer conference center. We got settled in and I remember Jeff, the author of this book, was taking time to map out where in the room he was going to interview people. He actually moved the chairs around to find the perfect spot for the other attendees to listen, take notes, and for the interviewee to feel comfortable.

The next day, during an interview, a customer leader said something that really sparked Jeff's curiosity. Jeff asked if he could take us to where that part of the work was getting done. We then got a two-hour tour of the customer's facility. I had never seen a *requirements* person make such a pivot during a meeting, but Jeff did, and it injected us and our customer with more enthusiasm about getting the research right.

The next few days were filled with empathetic interviews, lots of post-its, and a few value stream mapping exercises. Once we returned to our home offices and after a few days of crunching through the research we had collected, Jeff and I saw what needed to happen. We quickly went into design mode and wireframed out an application. What would have normally taken months of back-and-forth with the customer, we clocked in at just under two weeks. We designed a business system that would help the customer solve their unique business challenge.

After we briefed the customer, the only additional item they wanted was for the system to have a mobile sidecar application so that executives could access information on their phones. We wouldn't have been able to understand the problem and how to remediate the challenge if it wasn't for Jeff's well-honed skills in UX research and design thinking.

This story happened while Jeff and I were working for GE Aerospace. I was the CTO of Military Solutions and Jeff was the lead UX Designer in an external-facing software division. We have several stories like this and Jeff has decades of them. Early on in Jeff's tenure with us I realized that he has a unique ability to get a group of people to recognize the real problem, maneuver that group to promote ideas and then to find solutions that exceed their expectations. He has a collaborative, scientific and empathetic approach that gives customers (and users) a sense of relaxation, fact-fullness and empowerment during his UX design thinking process and workshops.

My journey around user experience started off in the *dot com* era where we had interface designers, information architects, content strategists, and a cadre of other folks to elicit requirements. However, all these folks were hyper focused on their specialties and quite often I, as a software professional, had to act as a mediator of these captured requirements. Fast forward to today and I see the following:

- **Too many tech projects are successful failures**: Many tech projects are completed correctly, but they don't meet the business's desires or outcomes. I call these projects "successful failures," and they're all too common in business today.

- **User experience (UX) can help projects succeed**: UX is a discipline that can greatly increase the chances of a project meeting or exceeding expectations.

- **UX has become more commonplace in recent years**: Experience and a learning mindset are what will set UX professionals apart.

I am delighted that Jeff has put this book together; Jeff's book is filled with deep insights into the UX discipline. It's not a high-level overview, but rather a comprehensive guide to everything you need to know to help projects and the businesses they support succeed. From user research to design thinking to analytics, Jeff covers it all. He also shares his own personal experiences, which gives you a unique perspective on the UX field. If you're serious about UX, then this book is a must-read. It's the perfect resource for anyone who wants to take their skills to the next level.

Travis Wissink

Sr. Director, GE Aerospace

Contributors

About the author

Jeff Hendrickson is an honors graduate of the FIT design school who began his international career designing clothing and textile lines – one with a former classmate that showed in NYC to global acclaim.

A downturn in the clothing industry saw him go back to school to study technology, and it's in technology that he's remained, working his way up from designing small websites to leading UX teams for enterprises spread across the globe. He's logged many miles on jets, crisscrossing the planet to teach design thinking and help companies discover new ways to serve users and customers.

Jeff currently lives in Cincinnati, OH, where he continues his work in the UX field, while he paints, publishes, teaches, and produces bartender cocktail competitions and art festivals under his company, Alchemy Entertainment Group.

About the reviewer

John Coria is an experienced design leader with 18+ years of experience in creating, facilitating, exploring, and implementing UX and UI concepts to drive results for customer-facing and technology-focused organizations. He has an extensive and proven track record of passionately reinventing the way customers use technology by leading and implementing design vision and strategies, articulating and presenting ideas to teams effectively, and working across organizational boundaries. He also enjoys building and managing high-performing and cohesive design teams by leading, developing, training, and providing effective coaching to attract and foster talent, while supporting a culture of collaboration, creativity, and innovation. As a roadmap builder, he is a visionary who is enthusiastic and flexible, with the ability to influence through conviction, thoughtful strategies, and a commitment to culturally aligned, value-focused, brand loyalty success, while setting new standards in executional and operational excellence.

Table of Contents

Part 3: Research

7

Design Thinking: Understanding the Core Problem-Solving Tool in a UX Practice   57

8

Interviews and Personas – Best Practices for Quick and Efficient User Understanding   71

12

The Tools It Takes to Gather and Consolidate Information   129

Part 4: Strategy

13

Digital Transformation – An Enterprise Use of UX Analytics   143

14

Content Strategy and Analysis – an Analytics-Based, User-Centric Approach   167

Part 5: Tactics

15

16

17

Preface

UX Analytics is a field that recognizes the significance of understanding human behavior and emotions when it comes to designing user experiences. It goes beyond mere metrics and embraces a people-centric approach. By delving into this book, you will acquire essential skills, knowledge, and techniques to establish a top-notch UX Analytics practice.

This comprehensive guide will equip you with the strategies and tactics required to effectively collect, analyze, and interpret data, enabling you to make informed decisions that enhance the overall user experience your customers expect. Additionally, it emphasizes the importance of empathy in comprehending user needs and desires, allowing you to create meaningful and impactful design solutions.

Additionally, this book will walk you through the entire UX Analytics process, from setting goals and defining **key performance indicators** (**KPIs**) to implementing various research methods and tools. You'll gain insights on user interview best practices, usability testing, and techniques for gathering qualitative and quantitative data. By mastering the art of data analysis and interpretation, you'll be able to uncover valuable patterns, trends, and user preferences, enabling you to make data-driven design decisions that lead to improved user satisfaction and business success.

Overall, this book aims to help readers effectively leverage analytics to improve user experience. While stressing the importance of data, it also covers potential limitations and pitfalls. The book offers an invaluable, comprehensive resource for implementing UX analytics to drive better design. Key takeaways include the fundamentals, repeatable processes, practical guidance, and real-world applications.

Who this book is for

This book is for product managers, UX researchers, designers, and anyone involved in UX and business development, both in management roles and as individual contributors. If you are looking to master the methodologies, principles, and best practices for driving product design decisions through UX analytics, this book is absolutely the right pick for you. While a basic understanding of user experience principles is beneficial, it is not a prerequisite, as everything you need to know will be explained.

What this book covers

Chapter 1, *Understanding the Fundamentals of User Experience*, explains the importance of knowing what UX really is, and what its not. This chapter will help readers understand the broad scope of UX to improve user acceptance and usability of a software product.

Chapter 2, Just What Do We Mean by Analytics?, talks about the need to use analytics to understand user behavior of a software product. By learning the meaning and different types of analytics, readers will be able to design the correct research and interpret the findings with greater accuracy.

Chapter 3, Using Analytics Effectively for Successful UX Projects, teaches the importance of analytics and that analytics relies on the correct type of research to bring them forth.

Chapter 4, Achieving Success Through a Repeatable Process, shows how to spin up new projects quickly and efficiently. This chapter teaches how repeatable processes are sustainable and allow others not directly involved to understand quickly and follow the process at a broader level. They also set the team up for success by providing the framework for future projects.

Chapter 5, Northstar Statement - Your Greenlit Marching Orders, dwells into creating the Northstar statement in collaboration with the cross-functional core team. It takes input from every facet of the business to ensure that the project will meet the goals of all, including, business, product, and engineering.

Chapter 6, Change Management - It's Not Just About the Systems, It's About the People, focuses on how a properly staffed and well-trained change management team will help ease in the disruptions that a new application will unintentionally cause. It teaches the readers to prepare the workforce for the change well in advance of the release by making training videos and holding short, guided practice sessions with the power users.

Chapter 7, Design Thinking - Understanding the Core Problem-Solving Tool in a UX Practice, takes you through step-by-step processes and best practices so that you can become a respected facilitator of design thinking.

Chapter 8, Interviews and Personas - Best Practices for Quick and Efficient User Understanding, explores the different levels of interviews to get and solidify the ask from management, then to test that ask against managers and individual contributors. Understanding the ask from business stakeholders and the problems or new goals from the bottom up and the top down is critical to a successful project outcome.

Chapter 9, Surveys - A High-Level, Broad-Stroke Tool for Quick Direction, solidifies the importance of moving out into the market and getting a broad sentiment of the problems you're trying to solve for. It explores the topic of surveys, which are a way to do this quickly and more efficiently than is often possible through in person interviews.

Chapter 10, UX Metrics - The Story of Operational Measurement, explores quantitative and qualitative metrics, understanding the ask, and determining the best type of metrics to gather and report.

Chapter 11, Live By Your Data, Die By Your Data, dwells on how a UX Analytics practitioner must become intimately familiar with all measurable aspects of their business so that they can support the goals of the management teams. It teaches the broad perspective of the data first, with an understanding of who it's important to and why.

Chapter 12, The Tools It Takes to Gather and Consolidate Information, gives you a brief understanding of some of the best tools available in the market - with links to the sites – so that you can do your own due diligence to determine what would be best fit for you, your company, and the work you do.

Chapter 14, Digital Transformation - An Enterprise Use of UX Analytics, covers best practices for digital transformation including change management, corporate sponsorship, and a strong, cross-functional core team of leaders to keep it moving.

Chapter 14, Content Strategy & Analysis - An Analytics Based User-Centric Approach, talks about how content is just as important as data and research. Content is the structure that is visible to users and therefore the face of the company. Hence, the capturing and storage of, and the access to content is a key aspect of digital business.

Chapter 15, Deliverables - You Wrapped the Research, Now What Do You Present?, ties everything together and teaches the importance of packaging it all as deliverables, that are agreed on by the core team, and can be adjusted based on analytics and other research as the iterative process moves through.

Chapter 16, Data Visualization - The Power of Visuals to Help with Cognition and Decisions, shows how by visualizing research and analytics, you can deliver information in a form that allows for quicker understanding and decision making.

Chapter 17, Heuristics - How We Measure Application Usability, delves into heuristic evaluations and their significance in the context of delivering effective UX Analytics outcomes. It helps study and learn usability heuristics so that designs can mirror most closely the way that humans expect applications to work.

Conventions used

There are several text conventions used throughout this book.

`Code in text`: Indicates code words in text, database table names, folder names, filenames, file extensions, pathnames, commands, and keywords. Here is an example: `B17_int_0825_shippingmngr1.docx`

Bold: Indicates a new term, an important word, or words that you see onscreen. For instance, words in menus or dialog boxes appear in **bold**. Here is an example: "Since this is a short sample of data, it's easy to see that in the **Sales Person** column, there are cells with no data."

> **Tips or important notes**
> Appear like this.

Get in touch

Feedback from our readers is always welcome.

General feedback: If you have questions about any aspect of this book, email us at customercare@packtpub.com and mention the book title in the subject of your message.

Errata: Although we have taken every care to ensure the accuracy of our content, mistakes do happen. If you have found a mistake in this book, we would be grateful if you would report this to us. Please visit www.packtpub.com/support/errata and fill in the form.

Piracy: If you come across any illegal copies of our works in any form on the internet, we would be grateful if you would provide us with the location address or website name. Please contact us at copyright@packtpub.com with a link to the material.

If you are interested in becoming an author: If there is a topic that you have expertise in and you are interested in either writing or contributing to a book, please visit authors.packtpub.com.

Share Your Thoughts

Once you've read *Unleashing the Power of UX Analytics*, we'd love to hear your thoughts! Scan the QR code below to go straight to the Amazon review page for this book and share your feedback.

https://packt.link/r/1804614742

Your review is important to us and the tech community and will help us make sure we're delivering excellent quality content.

Download a free PDF copy of this book

Thanks for purchasing this book!

Do you like to read on the go but are unable to carry your print books everywhere?

Is your eBook purchase not compatible with the device of your choice?

Don't worry, now with every Packt book you get a DRM-free PDF version of that book at no cost.

Read anywhere, any place, on any device. Search, copy, and paste code from your favorite technical books directly into your application.

The perks don't stop there, you can get exclusive access to discounts, newsletters, and great free content in your inbox daily

Follow these simple steps to get the benefits:

1. Scan the QR code or visit the link below

https://packt.link/free-ebook/9781804614747

2. Submit your proof of purchase
3. That's it! We'll send your free PDF and other benefits to your email directly

Part 1:
Introduction to UX Analytics

In order to carry out any **User Experience** (**UX**) project successfully, it's essential to have a clear understanding of what UX truly encompasses and what it does not. Misconceptions in the corporate and enterprise world often lead to unrealistic expectations that don't align with the core principles and direction of UX as a discipline focused on creating better experiences. Additionally, grasping the meaning of analytics plays a vital role in comprehending how and why analytics are applied in UX projects.

The significance of analytics cannot be overstated, as they rely on the right type of research to provide valuable insights. Furthermore, incorporating repeatable processes into UX projects is crucial for sustainability and success. These processes not only enable broader team understanding and streamlined collaboration but also establish a framework for future projects, setting the stage for continued achievements. This part has the following chapters:

- *Chapter 1, Understanding the Fundamentals of User Experience*
- *Chapter 2, Just What Do We Mean by Analytics?*
- *Chapter 3, Using Analytics Effectively for Successful UX Projects*
- *Chapter 4, Achieving Success through a Repeatable Process*

1

Understanding the Fundamentals of User Experience

Let's start off with one basic but extremely important understanding –**User Experience** (**UX**) does not equal **User Interface** (**UI**). UX encompasses a user's entire experience in interacting with your software product. They do this through the UI; the content, buttons, cards, links, and so on. If those are designed correctly and make the job of the user simple, easy, and intuitive, then the user has a great experience. In this context, UI supports and is part of the UX. The UI either allows for a great experience or it doesn't. And if it doesn't, then your job is to do the research and provide the analytics that uncover the current problems so that a solution can be devised and incorporated.

In this chapter, we will be covering the following topics:

- Getting familiar with UX

- Busting the myths about UX

- Understanding the differences between UI and UX

- Who does UX?

Getting familiar with UX

When we think about UX, we need to understand the breadth of the discipline. When we talk about it and evangelize it within companies that are new to UX – such as in a 1-2 on the maturity model – we need to always refer back to the user or the customer. Another key point to remember is that **Return on Investment** (**ROI**) must always be proven.

This will result in a focus on the direction of the product and engineering groups within the organization as well. By showing depth in the research, and if your team is involved, the solution discovery process findings, you'll be paving the way for a smoother full-cycle process of *UX > Product > Engineering*.

As a group of professionals who uncover or discover a hypothesis and then test its efficacy, UXers use methods and best practices to uncover the truth, which can be either positive or negative. Companies just starting down this path will incur expenses in management and individual contributor roles to start with.

So, what are the fundamentals of UX? In loose order of importance, they are research, design thinking, iteration, reporting, design, and testing. And while sprints are moving forward and we're getting in front of users to test designs and hypotheses, we'll consistently be watching for the universal criteria for good UX, as shown in the following figure (Credit: `Usability.gov`):

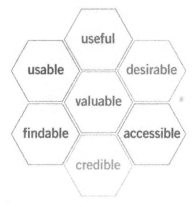

Figure 1.1 – UX honeycomb diagram

When I teach teams about the values shown in the preceding figure and use them in direct interviews with users, we have people rate them on a scale of 1 to 5, with 1 being the most difficult. We've had people ask if they can give 0 or at best 0.5! They dislike the current state of the software used in their work that much, and when we combine these numbers together for all interviews, we create a simple bar chart to show how each value scores. More about that later, in *Chapter 17, The Power of Visuals to Support Cognition and Decisions*.

If we go a bit further, UX is the ease of readability, the proper placement of a button, or the path from start to finish in a workflow. It's the empathy we practice with our users so that we get them to be honest and open with us about the problems and pains they're facing when using the software. UX is about finding what's not working and creating a solution to fix those problems. It's about bringing together the corporate view of a need with the contributor view to find the best possible go-forward plan.

The bottom line and key takeaway is that UX is strategy. It's the what and the why. We're going to do *this* and we're doing it because of this *reason*.

Busting the myths about UX

Now that we have a decent understanding of what UX is, it is equally important to understand what it isn't. UX isn't the pixels on the page. It isn't the code underneath or the decisions that a product team must make about how to provide the proper data to the people who need it. Yes, those decisions support the UX, but they aren't in the realm of the UX practitioner.

The UI isn't the UX either. While it supports the overall experience that a user has with your application, it relies on all the work done by the UX team. Wireframes are produced from research and design thinking sessions, and then tested with users to see whether the direction is correct. When there's a greenlight to move into the development cycle, the UI designers move into action and ask questions such as the following:

- Which design system are we using for this?
- What's the best card to use for this element?
- Is this font the right one?
- Can I place the button as specified? Can it be moved 20 pixels to the left and be approved?

Right? These people need to take what's given from the handoff and start creating the user interface that will drive the application. They're essentially working from a blueprint to build a house, car, or office building. They're asking questions when a design challenge occurs and working with a UXer to ensure that the usability, usefulness, accessibility, and so on are still according to the plan – the plan, as we now know, that's been formed from the research, design thinking, and iterations through the beginning stages of the project.

So, now let's circle back – these things aren't UX. UX sets the stage, but this is UI.

The bottom line and key takeaway from this section is that UI is *tactics*. It's the *how*.

Understanding the differences between UI and UX

In this section, we will see a point-by-point breakdown of the differences between UI and UX. Understanding this is vitally important to a strong UX practice within an organization, and you must be able to help others understand this as well. Let's look at these:

- **UI supports UX**: The user interface is what people interact with in an application. It's the buttons clicked, the body copy and titles, the white space, and so on. If these are designed correctly, and users understand how to move around and do work, then the UI has supported the UX the way it should.

- **UI and UX require different skill sets**: A UI designer knows how to code, a UX designer in general doesn't. UI designers can create CSS classes and know HTML and JavaScript. They're responsible for creating what the UX team has specified for the design of the application. A UX team says, "This is what needs to be designed," and the UI team does that work to tie the design into the system.

  A UX person learns and understands empathy and rapport. They work directly with the user and customer to uncover problems and to ensure that the new designs are satisfying demand and expectations. True UX is more about research and iterative design phases than it is about anything else. It's about doing the absolute best job of giving the UI team a solid start to build with, which will in turn smooth the process and allow for quicker transformation to a market-winning product.

- **UI is what an application looks like**: Briefly touched on earlier, what a person sees and uses is the UI – user interface. It's the colors, the sizes of the buttons, the amount of white space on the page, and the placement of titles and paragraphs. If the UI is consistent with the company brand, then it appears to belong to that brand. If blue *#6C9FB1* is your brand's blue, and the UI team uses *#317EB1*, the UI isn't correct and breaks the brand promise of the company.

- **UX is how it supports the needs of the user**: Go back to the meaning of UX – user experience. If the brand promise is kept, and we listened to the users and built what they said would make their lives easier, we're creating an experience for them that satisfies them and makes it easier to do their jobs efficiently.

Next, let's explore the roles that perform UX.

Who does UX?

With UX becoming one of the hottest career paths in the past few years, specialties such as UX architect, UX designer, UX researcher, UX strategist, UX writer, and more are being born. And while they all *do* UX, their roles differ. An architect or a strategist has the broadest work priorities, while a researcher, designer, or writer has a tighter set of responsibilities.

As I mentor people who are just getting started, I advocate being a full-stack UXer – having a core strength and building on that with knowing how to do any of the other parts. Since UX is still a relatively new discipline – especially in the US – some companies will just need to hire one or two people to start building a team around. Those people will be full stack UXers. A designer alone won't get the job nor will a researcher. If the company wants to figure it all out first, they may hire a strong strategist to start with plans, or an architect to start working with developers and engineers. A strategist that can design, do research, and write is one of the strongest UXers in the market – always.

UXers are curious by nature, I believe. They like to talk to people and help them solve problems, sometimes through rigorous and sometimes through soft inquiry. We're comfortable expressing our views and opinions and will always be the strongest advocate for the needs of the user and/or customer. We understand that building rapport with a user will always get us the best and most honest feedback, and therefore, the best results in the final product. We're okay with ambiguity because, at times, it's all we've got. We can find a common strand of information across several planes or channels of expertise, problems, needs, and goals.

Understanding the functioning of a UX Team

A tight UX team is a cross-functional unit where all of those roles mentioned earlier work together to accomplish the goals. A researcher will rely on a designer to help turn wants and desires into beginning-level, low-rez wireframes. A strategist will brief the entire team on a direction for an upcoming project and review all the expectations and deliverables. A writer will ensure that everything is covered in reports and other collateral, and in some companies, even write all the copy, instructions, and error messages that go into an application, and if they aren't directly responsible for those, they are advised and used as a final editor. An architect will be working closely with data and software people to ensure that the backend and frontend needs are understood and met. They will work through blockers with product and management to either pivot or figure out the best plans to get those blockers dissolved so progress can continue.

A tight UX team works well together and they work well in the collaboration that's needed with product, engineering, marketing (at times), business development, and sales. Those companies that score highest on the maturity model understand this and it's become the heart and core of the way the company does business. Challenges to thoughts and ideas have to be made but they're made with respect for the others on the team. Not everyone agrees on direction, strategy, or tactics, so, we often use a design thinking session to whiteboard the situation and work through it visually. Together, as a unit, we move forward, making positive changes in the company and providing great value to the corporation.

I've trained teams around the world that comprise many mental model types. Engineers, **Vice Presidents (VPs)**, tech managers, developers, designers, researchers, report writers, production line controllers, shipping clerks, and on and on. Some of them will use what they learned, and some won't. I've gotten emails from managers telling me how vital the training was to their team. I've watched VPs and CEOs walk out of training for a break still talking about how they could have asked that question during interview training better. Or even still, being in character of the role they played when learning the skills needed to use empathy, build rapport, and get down to the honest problems that someone is having.

It's quite remarkable to see changes within a corporation as they begin to embrace the principles of UX and support efforts to grow it to maturity for the benefit of all involved. A UX practice within a company can open opportunities they never knew existed. It can improve morale by pushing for better workflows and systems for higher productivity, and it can lead the company in the direction to cut inefficiencies and improve profits. In short, a strong UX team can be the driving force behind digital transformation to move a company to the next level.

To conclude, a UX practitioner is a curious, empathetic, and diligent seeker of problems to solve.

Summary

In this chapter, you learned how to tell the difference between UX and UI, and to identify what types of roles practice the UX discipline. You also learned the starting basics of what a UX team is and how they function within an organization.

These points are important to you because UX is gaining momentum as a mature and important practice in companies across the globe, and if you know this going into a new role or even being tasked with expanding UX within your organization, you'll be successful much more quickly than if you didn't have this core understanding.

In the next chapter, we begin to dive into the core fundamentals of UX analytics. We'll uncover the two main types of metrics we use to measure and assess current performance and users' usage patterns, while we gain an understanding of research readouts.

2

Just What Do We Mean by Analytics?

Knowing what's meant when we talk about analytics is another key to understanding how and why you're attempting to use analytics for your UX project. In *Chapter 1*, we learned to distinguish between UI and UX and to help others understand that as well. Having that knowledge is foundational to everything you do as a UX professional.

In this chapter, we'll explore several factors of UX analytics as the foundation for the remainder of topics that will be covered in further chapters.

Always keep in mind that a solid UX practice is ever evolving and always moving, existing within a constant improvement model that is very similar to the Kaizen discipline. UX is a job that's never finished, no matter the scope of the project. A team that's strongly tied to company goals and possibly **Objectives and Key Results (OKRs)** should be involved in all work being done and put out into the world for users or customers. The objectives part of OKR are the derived goals – that which is sought to come to fruition as a result of the work – and the key results are how the goals of the objectives are measured and reconciled.

In this chapter, we will be covering the following topics:

- What UX analytics entails
- Understanding the impact of analytics on successful UX projects
- Exploring common types of metrics

What UX analytics entails

Let's go back again for a minute to that definition of UX. It is the set of expectations that a user comes to an application with, and the measurement of how closely we meet or exceed those expectations. One of the surest ways for us to gauge how we're doing is to create a set of experience criteria that we'll test against users. The set that was shown in *Figure 1.1* in *Chapter 1* is a start, and those are *qualitative* metrics. By using, for instance, the 1 to 5 scale, we're testing the quality of our work by allowing people to do a digital task within a tool, series of tools, or a system.

So, the crux of UX analytics is our steps, practices, and reporting structures that either prove or disprove what we're trying to accomplish – in other words, testing our hypothesis. If we build products, we should start with rapid prototyping and user acceptance testing to keep steering us in the right direction. I've seen companies spend $500,000 on a product but not utilize UX research, and then burn through another $200,000 in development costs – that is, tech debt – only to have users not adopt it and keep doing jobs the same inefficient way they've been doing it for the last x years.

What we're testing for and measuring are the criteria by which we determine our direction. If it's a software product that's supposed to eliminate five steps of manual labor, then we start by measuring the amount of time it takes to do the tasks and how that then converts to time wasted and labor costs. Looked at from a practical viewpoint, this could be the reduction of technical and design debt, which increases operational and development efficiencies.

Let's take a look at a couple of use cases to better understand this.

Quantitative metrics example

Let's say that a manufacturing facility loses money due to inefficient inventory control systems. UX research is conducted and analytics are reported. Here is the scenario.

> **Scenario**
> There are three shipping clerks earning $50/hour x 4 hours/day = $3,600/week on a 7-day week, $14,400/month, and $172,800/year.

Research and analytics show that we can automate 2 parts of their workflow and tasks to eliminate 65% of that cost and save $112,320 a year. The automation will also speed up processes that will allow the company to expand to a new class of customers and generate additional revenue.

The cost of the automation will be regained in 1.75 years and provide net positive results thereafter.

The math then looks like this – `$172,800` - `$112,320` = `$60,480`, or 35% of the original cost.

A qualitative metrics example

Let's say that a healthcare facility loses patients to a competing hospital that's out of the way for most local residents.

> **Scenario**
> The hospital is staffed properly, but quality surveys are showing increasingly poor scores for the past two quarters. People aren't happy with the service they're receiving or the attitude of the staff.

On average, these problems cause the loss of 15 patients per month. These are measured against historical numbers for each month of the year, and adjusted for weather, accidents, and crime rates in the area served by the hospital.

Here are some problems I found while working on a project at a hospital in Ottawa, Canada, a few years ago:

- Missed medications (meds)
- Spills in halls not getting cleaned quickly
- Patients missing operation schedule windows
- Nurses not showing up for work on time
- Often impatient and rude staff
- Patients falling

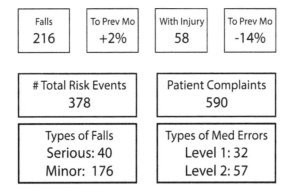

Figure 2.1 – A wireframe example for a dashboard

I'd wager to say that these are problems many hospitals around the world face daily. We use UX research to determine, first, what we need to measure in the processes – both manual and software system or tool-based – and then when metrics are analyzed, we can figure out what needs to change to eliminate the customer service problems.

The preceding scenario shows that when starting with poor qualitative numbers as the basis for the work to be done to improve the scores, we must look at the root causes of those problems. Sometimes, we prove that they're morale problems due to overworked employees, and other times, we prove that the current processes are inefficient for the workload and demands on the facility and staff.

Here's what a dashboard might look like for the qualitative measures if the hospital did surveys, rated on a 1–5 scale, with 1 being noticeably terrible and 5 being very good to excellent service.

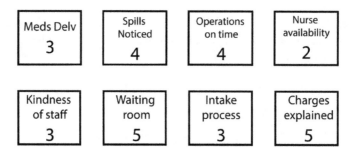

Figure 2.2 – A qualitative measures dashboard example

Okay, so we've been through qualitative and quantitative examples and have seen a few very simple dashboard graphic displays. Now let's get into what this means as far as an impact on a business and how they can improve where needed.

Understanding the impact of analytics on successful UX projects

At its very core, a UX project is initiated and managed to fix a problem of UX within a system or software product. Companies spend a lot of money on software, the engineers to implement the infrastructure to support it, the developers to code it to tie it into databases and possibly other current systems, and the UI designers to create the look and feel. There are also course designers whose job it is to create onboarding materials that will teach others how to use the new software, so all in, a company's financial outlay only starts with the initial purchase, but as you can see, it then involves many other expenses to get it up and running and then continue to run into the future.

All companies want to know how fast they will get their **Return on Investment** (**ROI**), and that's where UX analytics has a huge role to play.

If the proper work was done before a purchase, then the company knows that $*XXX* was being wasted each year on inefficient workflows and manual processes. Data isn't trustable and often resides in spreadsheets on someone's laptop, rather than in a database in the system. Often, these inefficiencies cause rework. An analyst gets paid $75 an hour, does six hours of work on a project, gives the results to the next person in the workflow, and 2 hours later, is told that the figures are off by 15%. $75 x 6 = $450 down the drain. Multiply that by 6 analysts, 12 nurses, 3 report writers, and so on.

UX analytics should start any software or digital transformation project. First, figure out what's being wasted or left on the table because of the current state of the inefficiencies and problems, and then quantify all that. Next, figure out the cost of a new system – the total cost, as we discussed previously – and estimate the savings realized by cleaner, more accessible data and simpler workflows that will save hours of rework. Then, make the purchase decision to buy or build a new software system.

Follow those guidelines and steps to successfully work through the problems and make the best choices. The upcoming chapters will show you how to do it all.

Exploring the common types of metrics

As alluded to earlier, the two types of metrics are as follows:

- Quantitative
- Qualitative

Let's look at both in a bit more detail.

Quantitative

These are ratings along a scale that show values. Some examples include the following:

- **Units**: For example, the number of handbags sold in the southern region of a department store
- **Revenue**: Money made from the sale of those handbags
- **Inventory**: How many bags of organic lemons are currently in stock
- **Salespeople**: How many salespeople earned bonuses over $XXX in the first quarter

These are numbers and objects that can be counted, divided, added, and multiplied. We'll learn how to visualize these in *Chapter 17*, so for now, put the definition in your memory bank – things you can count, add, do other math on, and so on.

Qualitative

These are metrics that show sentiment. Some examples include the following:

- **Happiness**: Did the customer service make the buyer happy?
- **Satisfaction**: Are you satisfied with your experience using the new weight loss app you just purchased?
- **Met expectations**: Were the expectations you came with to the new dance class met?
- **Worth the money**: Was the investment you just made in the vacuum cleaner worth the money?
- **Net promoter score**: This is the percentage derived from subtracting the negative from the positive to get a score.

Quantitative versus qualitative metrics

Here are a few quick examples to see the difference between the two metrics types:

- The number of apples in the bag versus confidence that the apples will be tasty

- Sweaters on sale on the front table versus customers that will find them appealing

- Commissions made by the top salesperson in a region versus the happiness of all the families that bought a house with them

It's important to point out here that both types of metrics are important to capture for most projects. Counting the number of clicks on a particular button in an application can also lead to a percentage of users who clicked it to get to whatever part of the application it leads to. If it's a section on rules and regulations, for instance, you now know what percentage of your users are interested in reading that type of information.

With this, we can wrap up this chapter with some key takeaways to remember as we go along this journey.

Summary

In this chapter, you learned what the function of UX analytics is and the two main types of metrics that give you the best analysis of performance for your applications:

- **Quantitative**: The metrics that give you numbers, such as sales volume, the number of students per class, or the cost of a new pair of shoes

- **Qualitative**: The measures of quality, such as ratings of service, happiness with a product, and expectations met on a cruise

When you know what direction you want to head in, you can choose the right metrics for your analysis.

Okay, so let's keep moving. Next, we'll dive into the *why* of analytics and how an understanding and the use of proper practices can set you up for success.

3

Using Analytics Effectively for Successful UX Projects

The importance of analytics can't be overstated and in the previous chapter, we learned about the two main types of metrics so that you always know what to look for, and how to combine them for the best results.

Analytics relies on the correct type of research to bring them forth. Just like when building a house or putting together a model car or plane, there must be instructions of some sort to get you the intended result. All successful software projects rely on the right UX analytics processes and metrics to build upon. UX analytics provide evidence of user behaviors, wants, and desires.

Knowing how users interact with your software products early and often as you progress allows for redesign if the findings aren't great and gives you confidence when you're doing things right. You can achieve this by utilizing a mix of two analytics types – quantitative and qualitative. This practice is always recommended for giving the best picture of user acceptance, whether it's poor or strong.

In this chapter, we will be covering the following topics:

- Knowing the goal before getting started
- Understanding how analytics is like a roadmap
- Mastering quantitative and qualitative analytics
- Showing the findings to the stakeholders

Knowing the goal before getting started

So, what's your direction? What are you trying to accomplish? How you go about answering those two questions is one of the keys to good UX analytics practices and a major part of this book and the methods intended for you to use in projects.

When we start a project – and this will be explained in detail in later chapters – we must understand where the stakeholder wants to go and then start heading in that direction. So, the very first iteration of the project is exactly that – the shared understanding that your team will work with the stakeholder to do the research and set the analytics models and tools in motion.

For now, this is the only understanding we need. We agree on a route – like on a roadmap – and we start from there, for example, *point A to point B will currently look like this*.

Here's what you should do each time to set a solid foundation for the project:

1. Grab the UX Project Stakeholder Blueprint.

2. Set a date to interview the stakeholder. If there are two or more, interview each one separately.

3. Get the team together and determine the primary interviewer.

4. In the interviews, follow the protocol (this will be covered in *Chapter 6*).

5. Create a soft copy master of the interview notes from each notetaker.

6. Go back to the stakeholder and write out the North Star statement that explains the intended outcome of the project.

7. Agree that this will be phase 1 of the process and that further interviews with users who will be involved in the project could change the trajectory.

8. Get to work on a plan to do the research and start the analysis processes.

Yes, this is extremely simplified, but right now, that's all we need. This is a summary and an overview of the lessons to come, so that you get a feel for the intricate work to be done.

> **Bottom line and key takeaway**
>
> You must understand the requirements and the direction must be derived from a clear agreement between the stakeholder and your research team.

Understanding how analytics is like a roadmap

And so, we view analytics as a pseudo roadmap. There are paths to be taken as we work through the project, guiding the dev team, possibly marketing, and sales – and always the stakeholder with whom you're working – to accomplish the goal.

There are many different paths, so what is the most efficient? Which will get us to the goal with the least amount of effort and cost? Which will cause problems in places we may not see? And what happens when we hit a roadblock?

With certain types of problems, we can actually figure out the best ways to handle the situation and then use that when that problem type arises again.

Let's use a run map as an example. We will look at two visuals from the app I use to explain this concept. There are two parts to it, the *wandering to find a cool route* part and the *this is the most efficient route to get where we need to get* part.

In the wandering phase, you're trying things. Go here, go there. In the software world, that will start with an evaluation of current states, such as staff to implement, database systems, APIs, security concerns, and so on.

In the most efficient route phase, you're pulling from past experience and use cases of projects, the collateral you created, and the best practices you noted to follow.

Let's go to the walk/run route map. The run map in *Figure 3.1* shows where I went and the distance I covered. It's a clear visualization of the route and can provide me with the means to do a similar walk in the future if I want to achieve the same distance:

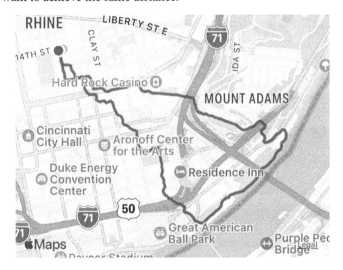

3.66mi Walk

Sunday, Jul 24 at 8:26 am

Figure 3.1 – Run map

I started the route with the idea that I wanted to walk along the Ohio river, go east, up through Mount Adams, and then back home.

Here are the statistics:

The following are a few quant questions:

- How many users logged into the finance app today?

- How many reports were created? How many were downloaded as PDFs?

- How much time did each user spend in the real estate section of the site, looking at single-family homes in the southeast?

- What were the sales figures for all stores in North Carolina for the first fiscal quarter of the year?

- How many airplane engines did we sell to American Airlines last year?

Then we have qualitative analysis metrics (quals), which are representations of, well, quality. This app doesn't account for that, but it's not built to do that, so I don't expect it. However, if at the end of the walk, as I was ending the session, it had asked me to rate my enjoyment of the walk, or how I felt physically during the walk, that would be another story. And here are a few qual questions:

- What is the average rating for our new class on analytics?

- How pleased were the patients with their nurses last week?

- How are the ratings for drive quality on the new Lexus model?

- Do people tend to favor our green paint or the gold paint?

- Do people with four kids want four bedrooms?

See the differences? Qual analytics can be yes or no answers. And they can also be numbers, but not numbers that are used for counting things. An average rating of anything is used as a guide for how you're doing. If the average rating for the new app you just designed is a 3 out of 10, you've got some work to do. We're looking for scales and ways to tell where we fall in acceptance.

"On a scale of 1-10, the overall satisfaction with the comfort of the new seats is a 6. Last year, with the older seats, it was a 4 so we're improving." A measured improvement shows the right direction for customer satisfaction.

Qual metrics are also what people use to judge whether or not to make most purchases on sites such as Amazon. They want social proof that a product is worth it. That's where star ratings come from. An average star rating of 4.3 means it's a good product and most likely worth the money. The caveat to that though is if only one person rated it. You can't really tell then, can you? But if 500 people gave ratings and the average is 4.3, then you can be confident it's a good buy.

We should also think of qual metrics as those that can have answers such as the following (which can be listed in a single select):

- Not satisfied at all

- Somewhat satisfied

- Satisfied

- Very satisfied

- Extremely satisfied

Or they can even be designed as a Likert sliding scale:

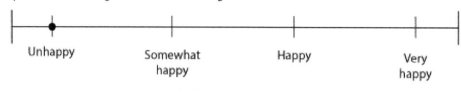

Figure 3.3 – Likert scale example

These answers are adjectives. "I'm *happy* with the experience," "I was very *excited* to find the new software," and "I was *disappointed* with my care at the walk-in clinic."

Figure 3.3 also shows you how to create a qualitative tool that doesn't allow for a non-conclusive answer. In this case, the users will tell you that they're unhappy or happy, but there's no middle ground. In many cases, giving a middle answer such as, perhaps, *neither happy nor unhappy* leaves room for people to just fly through your research without really thinking about how they feel. If you get back the results of that survey – with the middle answer – chances are you won't get the best results. Always consider this when using a Likert scale for research.

Quants are *how many* and *how much*. If people like the quality of our work, they will most likely be loyal customers. High quality usually equates to higher sales figures. But you don't get the bucks unless you're delivering that quality. Let the quality slip, fail to do the correct research up front, fail to keep promises, and dollars earned start to go down.

It's pretty much impossible to do one without the other. Both types of metrics provide the feedback and the direction a team needs to build the best products. The metrics from your analytics efforts provide evidence. And evidence leads a product team to either continue on the current path, shift a bit toward a slightly different goal, or stop and pivot to a new goal. Back up to that previous example. We know that 500 people – a quant metric – gave ratings of the product. If 500 people gave ratings and the average rating was 2.7, you'd probably hesitate to buy that product. Ratings are an excellent example of how quant and qual can be used together to determine direction, worth, value, and so on.

Bottom line and key takeaway

Both quant and qual are necessary to gather when doing UX analytics so that solid decision points are clear and actionable.

Now that you understand the metrics in even greater detail as we continue to break it all down from an overview to specifics, let's learn what we do with all the metrics we've gathered.

Doing the readout – showing stakeholders the findings

At the end of all this, you and your team will most likely be giving a readout to the stakeholders and the dev team. This will differ from company to company because of team structures and the maturity of the UX practice, but it's best to be prepared and understand best practices so that you're ready to deliver the best possible analysis, which, in a lot of UX analysis work, should include a confidence rating.

Understand too that the results of the research might not be what the development team wanted to hear. You may get back results that prove that the new changes or additions of additional tools to an application aren't wanted or needed by the users. In some cases, they might feel that it will complicate a smooth workflow that exists, and cause them extra work without providing value.

> **Key takeaway**
>
> Always be honest, transparent, and objective. Don't sway any findings in one way or the other to satisfy what others may feel is necessary.

Later chapters will get into more specifics, but the guidelines are simple:

- Using visuals is better

- Use a story structure to explain data

- Be succinct and not verbose

- Be prepared to back up everything with more data

- Involve design, development, and engineering

Let's explore each of these in detail.

Using visuals is better

The use of data visualization can't be over-emphasized in this regard. We humans are visual creatures and can make better decisions faster when presented with pictures rather than words. We can pick a slightly different color or shade of one item from a set of many. We can recognize size variations quickly and we can recognize groupings of like items with ease.

Use bar charts to show quick comparisons of one set of values within the same category, such as revenue numbers for all bedding departments in all the stores in the northeast.

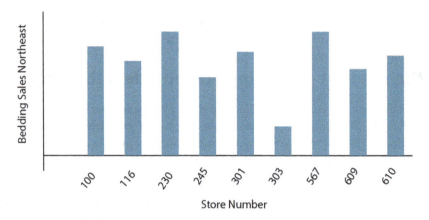

Figure 3.4 – Bar chart example

Use scatter plots to show comparisons of two groups, such as cash versus the net income of three Asian countries compared to the United States.

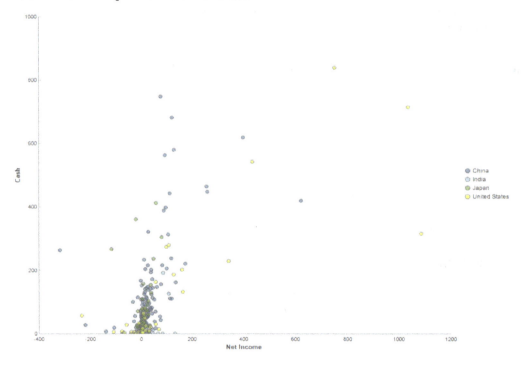

Figure 3.5 – Scatter plot example

Key Performance Indicators (KPIs) can be shown on a manager-level dashboard but make it *big*:

Sales this quarter	Losses this quarter
$42M	$2.4M

Table 3.1 – KPIs

A manager or vice president can open the app, glance at this, and know whether the company is on track or not. It's quick and easy and allows for fast decision-making either way. If the losses are average and acceptable when compared to the sales numbers, then everything is fine. But, if those losses are excessive and above average, there's research to be done and actions to be taken.

Using a story structure to explain data

Create a rich narrative around the analytics *story* you need to tell. Nothing is drier than just reading off a bunch of numbers. Use your personas to paint a picture of the users and their frustrations, or happiness and delight with the products and the workflows they use to get their jobs done. Be descriptive and make the readout interesting for people to listen to. It will make their jobs easier because they'll remember how you told the story more than they'll remember the numbers you showed them. This is all part of the empathy angle of user experience design.

Being succinct and not verbose

Yes, I just told you to tell a story and be descriptive, but that doesn't mean you have to wind on and on. Being succinct means that you stick to representing the numbers in a way that makes it clear to everyone what the research showed.

Being prepared to back up everything with more data

You always need to have more data in your back pocket. What I mean here is that while you may show something such as an analytics KPI, you may not show all the math and numbers behind that. But you should have all those close at hand for when somebody asks a question, and you need to get down to a finer level of granularity.

"Yes, Joan. The ratio of X to Y is 14% higher with the addition of the new branding colors. Interestingly enough, that figure is most consistent with users in the 20 to 25 age bracket, across all education levels and demographic areas."

Get the picture?

Involving design, development, and engineering teams

This is one of the core principles of Design Thinking – a cross-functional team. You don't want to leave anything from your analytics exercise to be misunderstood, explained by someone else, or inferred later during critical periods of work to make changes or institute new features and so on. You want everyone involved in the work to see, hear, and understand what you're presenting.

This collaborative way to work will help tremendously with the validation of all your findings because each of the people on your cross-functional team will have ideas and expertise to support the work and aspects of the work that need to be highlighted and clarified for understanding and quick decision-making.

Later chapters will give you examples and best practices of all the collateral you'll need, so be sure to download them all to start your library.

Summary

In this chapter, you learned that doing a good readout for your teams is just as important as the work you did to get the analytics. If you can't show it with impact, you may cause yourself – and the team – extra work.

You also learned the important steps in getting a UX analytics project started once you've been given the task, the order in which you'll carry out the best practices, the materials needed, the people to speak to, and so forth.

And to further strengthen your understanding of the types of metrics to gather, we went deeper into qualitative and quantitative analysis and how to best synthesize and present the findings from your analysis.

Next up, in *Chapter 4*, we're going to get into the importance of working with repeatable processes. This speeds up work tremendously and also ensures that the onboarding of new UXers to your team is smooth and efficient.

4

Achieving Success through a Repeatable Process

Any time you can create something that's easily repeatable, you should strive to do that, and do it well. Think about things that you do that you don't even have to think about anymore – tying your shoes, brushing your teeth, or making coffee or tea in the morning. You developed a level of mastery over those or other actions that now allows them to be done on autopilot, so to speak.

This is the same with a repeatable process in UX analytics, or any other discipline for that matter. You want to establish the rules and procedures as soon as possible so that there's very little thought process involved in getting started. Granted, once you get into the research, there will be deeper work, but you know that you're following a repeatable process so that the outcomes for projects 1 to 26 will all look and feel the same.

This also serves to create expectations for the groups for whom you're doing the work, allowing them to digest your work faster and with greater understanding, as well as knowing how and when to engage you and your team for the best and quickest results.

In this chapter, we will explore the following topics:

- Appreciating the beauty of a repeatable process
- Understanding how people think
- Understanding the importance of a strong foundation
- The importance of documentation

Discovering the power of a repeatable process

Repeatable processes create opportunities for sustainability and efficiency. They also allow for quicker onboarding of new team members if they're simple and easy to follow. A complicated process, while being very useful to some, will cause more people to ignore it than use it. This holds true for the processes needed to complete meaningful analytics, just as it holds true for workflows within an application.

When considering UX analytics, this is especially true for success. And part of the magic is that the process to create a repeatable process is a repeatable process. I'll explain via some examples and diagrams, but first, let's look at the parts of a UX analytics process:

- People and roles

- Data

- Tools and applications

- Processes

Let's take a deeper look into each of these.

People and roles

How is it that 20-year-old females can understand your application's workflow 15% better than 45-year-old males? Why is it that children in one part of the world make use of education software 65% more than children in other parts of the world? And why do middle managers in manufacturing businesses prefer brand X warehouse management software over your company's very similar product?

People and roles are extremely important to understand because, until everything in this world is automated, people still work in the production of everything, from this book that you're reading to the satellites circling the globe. And it's those same people that will either enjoy working with your software or dislike it so much that they refuse to use it, because of the poor design and how much time it takes from their day to figure out how to use it.

For the past 20+ years, companies have been putting me on planes and sending me to places to figure out why people won't use software. The processes that I'm teaching in this book are the same ones I use with every company I visit. It's a repeatable process, and although it took me a few years to perfect, it's served to secure multimillion-dollar contracts and saved companies hundreds of millions of dollars.

People are the heart and soul of any company. Many boldly state that in their company mission statements and core values. If you, as a UX professional, can understand what they mean when they make those claims, you can more strongly align with them because those values are our exact values in UX.

Empathy, nonjudgment, safe spaces – these are all people-centric core values that we practice every day. And just because part of what we do in UX analytics is automated with tools to gather information doesn't mean we don't continually support and strengthen those core values.

Let's consider two different role types to understand a bit more. We'll look at the sales manager of a large retail chain, and we'll look at an inventory clerk in one of the stores. We will concentrate on the research part of the process. We know what the interface is for the manager, and we know what it is for the inventory clerk. The manager can see what the inventory clerk sees but not the other way around. The clerk has a view limited only to their store and what they receive.

The beauty of the process is that we design the interviews so that when we talk to the manager, we're getting information on the people that report up through them. Each store manager is a direct report, and the clerk reports directly to the store manager. We interview both the sales manager and the store manager as stakeholders. While they have different overall goals, each is responsible for sales at the store level, so we treat them the same as far as the interview process goes.

The store manager is also very familiar with what their inventory people do to keep the shelves stocked, so we get closer to the clerk through the store manager than we do through the sales manager.

So, let's back up a little higher again; that sales manager has eight stores, so eight store managers. Each store has two inventory clerks, so there's a total of 16 inventory clerks that report to the sales manager. You'll use the exact same process, with the same stakeholder forms and questions to interview each store manager, and then you'll use the user interview form to talk to each of the 16 clerks. There's no guesswork involved. The foundation of your research will be solid because you have comparable metrics.

The results from each store manager interview will be condensed into a master store manager report, and the results from the 16 clerks will be handled the same. You'll have a master that represents the *Store Inventory Clerk Persona*.

In the processes as laid out here, you asked each store manager the exact same questions. You then asked each clerk the exact same set of questions. They're unique and pertain to their work, so understand that these are two different sets of questions – stakeholder questions and store-employee questions. It's also important to point out that you could get very different answers to the questions within each group. Three different stakeholders could give you three different answers, and that's okay because it allows you to see with more clarity that your current project task could involve bigger underlying problems.

Data

Over the past five years or so, companies have started to wake up to the reality of just how important data is to them. I've done research proving that 35% to 60% of corporate knowledge workers don't trust the data they receive to do their jobs. Because of corporate silos and the consistent trend of using spreadsheets and keeping your data to yourself, they question accuracy and trustworthiness. Other research has shown that roughly 45% of respondents have said that it can take them half a day to check data for accuracy.

Scenario

Imagine you're senior VP of JKL Company, and you're informed that the managers responsible for giving you reports every Tuesday and Thursday morning at 9:15 AM are spending almost half their day wasting time trying to figure out where the most accurate data is. They fear losing their jobs because of mistakes that could be made. Those mistakes can cause you – the senior VP – to make the wrong decisions, and those wrong decisions could cost the company millions of dollars.

As you can see, while people are the heart and soul of a company, data is the lifeblood. With low accuracy and poor confidence from workers, it can ruin chances of winning new business and, in many cases, will cause the loss of current business, as customers flock to other suppliers that can support them faster and more efficiently.

The repeatable process here is that you'll be asking everyone you interview for examples of their data. Most will give you spreadsheets that they use, and if you interview several people with the same job function, such as customer service reps for your product, each covering different regions or territories or even different levels of customers, you'll now have data examples to start doing some initial data visualizations with.

Tools and applications

We have people and we have data. What do the people use and how is the data stored and presented for use – tools and applications? If there are multiple tools that a user must wrangle every day to complete their tasks, that can be inefficient.

Case in point – a scheduler in an aircraft engine repair facility who must open six different applications in order to find everything needed to understand incoming engines and the repairs needed.

Do I have all the parts needed to make the repairs and get the engines back out to the customer? Do we have the repair staff needed to handle the load? Are there any special circumstances for this block of engines coming in?

In a scenario like this – and it's a scenario I continue to see as I travel and work with customers – it's easy to make mistakes because of the complications involved in jumping back and forth between multiple applications.

Many companies have older legacy systems with data structures that aren't compatible with current technology. You need to understand that and map it so that you can show how upstream and downstream applications either work together, or don't. Either way, you're diving into an enterprise-level understanding of all software currently being used, and how it all relates to everything else. You're getting the big picture of the enterprise.

Processes

To wrap it up, we must look at the current processes. We want to understand what it takes for data to get to people and in what applications it will be presented. We want to understand where the silos and gaps are, who's responsible for moving and managing data, what the compatibility is between the different applications, and the steps involved in each task.

Because the processes involve the other parts of people, data, and applications, we can create flow maps of every touch point from start to finish of a task. Those tasks can be part of a larger process or cover the entire process. It's here where we begin to understand dependencies and what other parts are reliant on either downstream or upstream decisions, data, agreements, and so on.

Think of the process as a strategy, and people, data, and tools as tactics. The tactics live in support of the strategy, and the overall process is fed by people, data, and the applications.

Understanding how people think

While a full discussion of how people think and learn, and the mental models that describe the types, are beyond the scope of this book, it's important to understand the basics so that you can do the best research and gather impactful metrics in your analytics efforts.

People tend to fall into just a few categories of thinking styles, which are as follows:

- **Auditory**: These people learn best when they listen and hear information
- **Visual**: These people learn best when they see things, such as demonstrations or how-to videos
- **Kinesthetic**: These people are *touch* learners, who pick up information faster by doing something
- **Read/write**: These people learn best by reading things and taking notes

Clues that have been gleaned from years of study on these learning types come from how people verbalize questions and answer questions asked by someone else. For instance, if I ask you whether you understand what I've just explained or shown you, and your response is, "Yep, I see," it can be inferred that you're a visual learner. If you were asking me whether I understood something, and you said to me, "You hear me?", that would indicate that you're an auditory learner.

So, now translate this to your work with analytics. Remember that a big part of the task is talking to people, not just running a program to find out how many times a button was clicked on. The perfect scenario is that you combine the two types of metrics. For instance, if you interview 10 users and 6 of them said that the body copy of the application was hard to understand, you may have read/write learners, and this may start to explain why a certain page was abandoned 65% of the time.

One method supports the other. If you had the metrics first, you'd go to the interview notes and find the supporting information, which should now lead you and your team to a path toward a solution. In this case, you need to rewrite some body copy to make it more understandable to a wider audience – that is, people with varying learning styles.

If you found, for instance, that a few of your people who've expressed difficulties with an application are visual learners, and you know that visuals aren't used in the design, then you'd infer that adding more visuals would help these learning types. And doing so won't hinder the way that the other styles use the application. You now have a partial solution that could immediately increase the usage of the tool. So, you get the additions made and test again to prove, or disprove, the point.

In this way, running through each of your users' learning styles and matching them to the problems you're hearing, you can begin to get better and better ideas for possible solutions, and ways to alleviate pains and challenges. Sometimes, these are very simple to implement, so put the information together and work through it with your team.

Now that we have an understanding of the core fundamentals, the inner workings of people, and parts that make up successful projects, let's look at the foundation upon which it all should be built.

Laying the groundwork – the importance of a strong foundation

The foundation is what you return to for each project, subproject, or task. As mentioned previously, where you start is always a function of the task at hand; sometimes that's just figuring out why users are not satisfied with the interaction between point A and point B, while other times, it could be a market fit project for a new offering.

Whatever the case, each has a foundation where roughly 60% is the same for each, and 40% is geared toward that project type. Think of the old adage, "We don't need to reinvent the wheel."

What's true about that is that every wheel is a circle and that it attaches to a car, bike, motorcycle, baby carriage, train, plane, bus, and so on, and the wheel allows those vehicles to move along the ground. So, if you're building something that needs to move along the ground, a wheel must be included.

For your purposes, the foundation is the four essential parts – people, processes, data, and systems. What will differ are the types of users, the process flows, the types of data, and the systems being used. When you account for all of these – for every project – you will have the 60% covered, and you can adjust the 40% to account for the differences. If you start with the big picture first to set the tone and the perceived direction by interviewing the stakeholder, you can then know which modules to pull from for the rest.

When I talk about modules, it's because this can be viewed as a modular practice. Just like tokens or CSS in code, or the blueprint questions we ask every person we interview, UX analytics modules are essentially plug-and-play parts that create the foundation and support our repeatable process parts toward success.

Next, let's move to the finishing touches, and the crucial part of any project that must be accomplished before handoff or readout to the next team responsible for work – documentation.

Documenting it all

Everything you do is incomplete without documentation. Period. Working in a void and not capturing all of your work in an accessible folder and format for everyone to reference is a hard path to failure.

Documentation has many forms – whether text documents, spreadsheets, photos, screencaps, bar charts, and other data visualizations, notes, interview blueprints, or audio and video files. All this contributes to the project as verifiable proof of work completed, metrics captured, people interviewed, and solutions suggested. All of this provides continuity for the project while reinforcing the rigor and process used to attain and process the information.

And all this – as hinted – goes into a folder for anyone to access, depending on your company's policies, of course. There may be sensitive information contained that only certain people will be permitted to see and use. Be sure to check your security and permissions policies before giving access to anyone. It's quite possible that you'll be given a space with a security model already in place, but don't take that for granted – check and make sure.

So, think back to the learning styles – auditory, visual, kinesthetic, and read/write. If you've gathered information from the examples given, everyone will get something useful. I have a colleague who can watch hours of videos of interviews, but I can't do that. I'd much rather read through notes because I know what I'm looking for.

Conversely, I'm also a very visual person, so I want to see data visualizations that support the research. Show me a donut chart of the percentages of anything we're measuring, or a line chart that shows a measure across a month or year, maybe subdivided by region or a subcategory that makes the most sense for the task.

There's one last key need and best practice for documentation, and that's proper editing. Let's face it, some of us are better writers than others. We use *then* when the proper word is *than*. We use slang that might not be understood by others. And we misspell words and don't use spellcheck. Having a good editor is a must for everything you put out in written form. Have a fact checker too. Make sure that what you produce is the absolute best it can be and you'll pile up great successes constantly.

Summary

In this chapter, you learned about the steps of a repeatable process, the learning types, the importance of a foundation to build from, and the necessity of sound documentation.

If you remember that UX analytics is more than just the metrics you get from a measuring tool that's watching your application, you'll always produce the best information by which to make the best decisions.

What you do as a UX-er tasked with producing analytics is one of the most important functions of a modern company – one that's embraced digital transformation or is already ahead of its peers in the digital frontier. Get these fundamentals right and you'll become adept at jumping from project to project if needed or leading the way in longer-term work.

In the next chapter, we will dive deep into writing a guiding document for the team and stakeholders – the North Star statement.

Part 2: Goals

This part delves into the crucial aspect of setting goals in UX projects and understanding the dynamics surrounding them. A well-defined North Star statement serves as the guiding light, crafted collaboratively by the core cross-functional team to establish the project's direction and goals. This statement acts as a constant reference point, allowing teams to continuously evaluate and align their work with research and evidence for accuracy.

However, it's important to acknowledge the innate resistance to change within the human race. To navigate through the disruptions that a new application might unintentionally cause, having a properly staffed and well-trained change management team becomes imperative. They play a pivotal role in easing the transition and minimizing resistance. The team's responsibilities include developing comprehensive change management strategies, engaging with stakeholders, and fostering a supportive environment to facilitate adoption and acceptance. With their guidance, the organization can proactively address concerns, provide necessary resources, and communicate the benefits of the new application, thus minimizing resistance and maximizing the successful implementation of the UX project.

In this part, we explore the significance of these components in effectively defining and achieving goals within UX projects, ultimately driving meaningful and impactful outcomes.

This part has the following chapters:

- *Chapter 5, North Star Statement – Your Greenlit Marching Orders*
- *Chapter 6, Change Management: It's Not Just about the Systems, It's about the People*

North Star Statement – Your Greenlit Marching Orders

The North Star statement serves as a guiding principle for UX analytics research and product development and helps align all team members around a shared vision. It encapsulates the core purpose and mission of the project, outlining what success looks like and how it will be measured.

By setting a clear North Star statement, teams can focus on delivering features and experiences that are aligned with the product's overall strategy without getting bogged down by minor details that don't contribute to the larger picture. This approach ensures that the team remains focused on the most critical aspects of the product and doesn't lose sight of its ultimate goal.

Important to understand is how a well-defined North Star statement helps with decision-making throughout the iterative research and development process. Whenever the team needs to make a decision, they can refer back to the North Star statement to see whether their choices align with the project's overall direction. This approach ensures that everyone is on the same page and working toward the same goal, reducing confusion or fuzzy direction, and minimizing costly rework.

Overall, the North Star statement plays a crucial role in the success of any project. It helps keep the team focused on the big picture, facilitates decision-making, and ensures that the project is heading in the right direction. In this chapter, we'll delve deeper into the North Star statement and explore how it can be used to drive effective UX analytics.

Here are the top three lessons that you'll learn:

- The North Star statement writing formula
- Testing the statement for feasibility
- The need for iteration by research, development, and engineering teams

The North Star statement writing formula

I first learned the method for writing the North Star statement at a leadership seminar weekend. It starts with a well-known formula called **CPR – context, purpose, results** – which is a commonly used framework for structuring presentations, reports, and, in our world as UXers, North Star statements! I teach this everywhere I go, and if you get familiar with it, you'll be able to use it not just for your projects but also for your personal life as well.

After these parts are all written, we will move into the final part I call the "so that" statement, and this will be our North Star. You do the work on the CPR to get what's needed to focus correctly on the work to be done, and the North Star keeps everyone aligned.

So first, let's break down the segments – context, purpose, results – as you will find them described in any search you do online. They vary slightly, but here's a good example of what you'll find to get you started. After you read through these, I'll show you how I learned the process and now teach it. This is because it helps you understand how to handle work requests and how the formula supports the project.

Segment definitions for CPR

The first element of the CPR formula is *context*, which refers to the frame of reference for your project. It's what some companies I've worked with use as the *mission* of the project. Context sets the stage for the information that follows and helps the audience understand why the project and the work being done are relevant. This element is especially important when presenting complex or unfamiliar information because it helps to provide a common understanding of the project.

The second element of the CPR formula is *purpose*, which outlines the reason *why* the project is being undertaken. The purpose articulates the goals or objectives of the project and should be clear, concise, specific, and measurable. It also helps justify the investment being made to successfully complete the project. This purpose element is what we'll use to write the North Star statement.

Finally, the *results* element of the CPR formula refers to the outcome or impact of the project. Results provide evidence that the purpose of the project has been achieved and demonstrate the value of the UX analytics research, the development time and costs, and the end results achieved. This element is crucial as it enables the business, stakeholders, and other management levels to assess the effectiveness of the project and its impact on the desired outcome, and the new or improved behaviors we're trying to influence or change completely.

Implementing a logical CPR process

Whether your team is tasked with creating a new application or enhancing an existing one, you're given direction based on results, not context. This tends to come to you from the business and is captured and understood further in the stakeholder interview. They're addressed based on shortfalls within the business, or new revenue or customer service goals that someone has determined need to be achieved.

An example is, "We're losing customers because our self-service portal is too difficult to understand and use so we need to make a better portal." We'll lay out this scenario in *Chapter 13* in the section on quantifying your data for the readout report. Here, we'll start with a list of results we need to achieve – and this is how we start our CPR, from the granular up to the broad context.

We gather all the intended results first, then work up to what the purpose of those results will be, and then we set the context. While this may seem confusing, you'll understand once you see the process and practice it with a few sample scenarios I'll provide for you to work on.

Step 1 – writing the results that need to be achieved

Enhancements to the customer service portal

The results to achieve and metric types are as follows:

- Regain lost customer business – quantitative
- Increase revenue by $175,000 per month – quantitative
- Regain customer loyalty – qualitative
- Regain customer trust – qualitative
- Have an easier-to-use portal – qualitative
- Have a portal that reduces the work our customer service reps have to do – qualitative and quantitative
- Reduce the number of incoming phone calls – qualitative and quantitative
- Show customers our market leadership – qualitative
- Help customers gain more business – quantitative
- Streamline workflows for customers and us – qualitative and quantitative
- Break into new markets – quantitative

There could be more, but we'll stop here and give you a couple more scenarios to work on at the end of this section. We can see that these cover both quantitive and qualitative metrics and we know that in this scenario, making the portal easier to use for internal users and external customers will increase our revenue, while also helping them with using it. Remember that customers will sometimes pay more for an easier-to-use service and workflow, no matter how much they recognize that your products are better.

Step 2 – turning the results into the project purpose

Now we need to take all these results and turn them into a purpose statement. This can be accomplished in several ways. One of the most effective ways when working with a team is to do a ranking exercise to determine the importance of the intended results. It's a great opportunity to use design thinking with the team. If you're working in an office, write all the results statements on stickies and put them on the wall. If you're doing it virtually, it will be similar, just in **Miro**, **Mural**, or whatever application your company uses.

Now use colored dots to have everyone vote. You can pick your colors; for example, use green dots for the results that each person thinks are a top priority, blue for the next top priority, and yellow for bottom priority. Since these are results that have been written as important, there will be little difference between green and yellow, but the ranking will ensure that you write the most powerful purpose statement.

Working from the list that we saw earlier, let's imagine that these four got the most votes:

- Regain lost customer business – quantitative

- Increase revenue by $175,000 per month – quantitative

- Have an easier-to-use portal – qualitative

- Streamline workflows for customers and us – qualitative and quantitative

These are the ones we're now going to write the purpose statement from. Here's a first pass you and your team might come up with:

The purpose of this project is to implement strategies that will increase revenue by $175,000 per month and regain lost customer business. Additionally, we intend to improve the user experience through the design and development of an easier-to-use portal that streamlines workflows for both customers and our organization, with a focus on improving overall efficiency and customer satisfaction.

I think that sounds pretty good, but what might be interesting is if you and your team take those four results and write a purpose statement on your own. Or use all of those bullet points (results), vote on them, and then write a new statement. And of course, if you're at this point in your own project, yep, you got it – write the results and then the purpose to get yourself into the practice quickly.

Step 3 – boiling the purpose down to the context

This part can be the trickiest because context can be an elusive target when there are people who have different roles and agendas on the team. While that's what you need, care must be taken to listen to everyone and practice nonjudgment. The context is your overarching statement about the goals to be achieved in the project and can be the best way to evangelize the work upstream.

Ensuring that your stakeholders agree with the context is important as well because the context is the best way to phrase the strategy to company management, if they haven't been closely involved, which in most cases they won't be. So, here we go. Working from our purpose statement, we again want to pull the strongest parts out of it and write a sentence, just one sentence, that encapsulates all the results we voted on. Maybe it goes like this:

Improve revenue and regain lost customers with excellent user experience, streamlined workflows, and a focus on efficiency and customer satisfaction.

How does that sound? How might you and your team improve it? Do you think it should include the revenue goal of $175,000? If so, how would you work that in so that it still stays concise and to the point? Think about that and write a few on your own for practice, which is really the best way to master this powerful technique.

Our complete CPR

- **Context**: Improve revenue and regain lost customers with excellent UX, a streamlined workflow, and a focus on efficiency and customer satisfaction.

- **Purpose**: The purpose of this project is to implement strategies that will increase revenue by $175,000 per month and regain lost customer business. Additionally, we aim to improve the UX through the design and development of an easier-to-use portal that streamlines workflows for both customers and our organization, with a focus on improving overall efficiency and customer satisfaction.

- **Results**:

  - Critical:

    - Regain lost customer business – quantitative

    - Increase revenue by $175,000 per month – quantitative

    - Have an easier-to-use portal – qualitative

    - Streamlined workflows for customers and us – qualitative and quantitative

  - Important:

    - Regain customer loyalty – qualitative

    - Regain customer trust – qualitative

    - Have a portal that reduces the work our customer service reps have to do – qualitative and quantitative

    - Reduce the number of incoming phone calls – qualitative and quantitative

    - Show customers our market leadership – qualitative

    - Help customers gain more business – quantitative

    - Break into new markets – quantitative

You can think of context as being like a mission statement for your project. It should be something that supports the overall objectives of the enterprise and lets your teams, management, and customers, if you're working with a cross-functional team of your people and theirs, know what they can expect, without knowing all the end results you wrote for the CPR.

Recall back a few paragraphs to where it was stated that the purpose is what we'd use to write the North Star "so that" statement. That's another very simple formula that goes like this:

We will *<do something>* for *<some group>* so that they *<can achieve something>*.

The "so that" is the most important part because it supports the purpose and will include the most important results, just like the purpose statement does. Let's keep with our scenario here and write it.

Here's that purpose statement:

The purpose of this project is to implement strategies that will increase revenue by $175,000 per month and regain lost customer business. Additionally, we aim to improve the user experience through the design and development of an easier-to-use portal that streamlines workflows for both customers and our organization, with a focus on improving overall efficiency and customer satisfaction.

So, our North Star statement could be as follows:

We will design and build a new customer service portal that streamlines workflows for both customers and our organization so that all users can focus on efficiencies and we can increase customer satisfaction and increase our revenue by $175,000 per month.

Here's another opportunity for you and your team to practice. Write several different versions and vote on which is the best. If you work individually, having each team member write their own version, you can then do a mashup of them all and get the best version.

Testing the statement for feasibility

Any North Star statement should be tested against the criteria of budget, capabilities, and value. Your lens will be the intended results from the CPR, and this is again where the cross-functional team comes into play. Where you have to now filter is in regard to the work to be done – the work that the team can actually do.

For example, while the team can create an easier-to-use workflow (presumably), they can't guarantee that the customer will become loyal to the company again. Right? Making work easier to accomplish should do that, but that's the only thing that the team can do. Hopefully, the result will be regained loyalty.

So, out of our list, which of those results will be actionable for our team?

- Regain lost customer business – no
- Increase revenue by $175,000 per month – no
- Have an easier-to-use portal – yes
- Streamline workflows for customers and us – yes
- Regain customer loyalty – no
- Regain customer trust – no

- Have a portal that reduces the work our customer service reps have to do – yes

- Reduce the number of incoming phone calls – no

- Show customers our market leadership – no

- Help customers gain more business – no

- Break into new markets – no

All the *no* bullets are dependent on many factors, including how well the new system is designed and built. The *yes* bullets are actionable tasks that the teams can undertake and deliver. If the team delivers streamlined workflows in an easier-to-use portal, then there's a chance that the other factors will be realized – but of course, none of those are guaranteed.

So, the feasibility study will only be concerned with what can be built or not built. Let's look at a matrix:

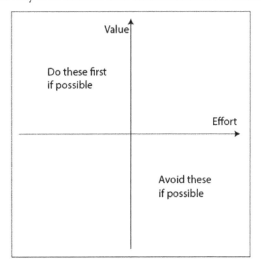

Figure 5.1 – Feasibility matrix

The top-left quadrant is where your analytics work found problems to be solved that are both high-value and low-effort, so if feasible, they should be done right away. They are feasible if your company has the tech and the budget to do them. If, for instance, the value stayed high, but for whatever reason you would have to purchase, let's say, a new server, then the effort via budget is a little higher and that would float to the right in the matrix.

This is a team effort because all parts of the tech team need to give input on the effort. The engineering team knows what it will take from the backend, while the development team knows what it will take to ensure that the interface meets all the needs for the enhancements. Anything relegated to the bottom right – high-effort/low-value – can be put on the back burner and revisited at a later date, perhaps when backend investments have been made, for instance.

A feasibility matrix can be a valuable tool in testing the results intended for a project, especially in relation to the goals and metrics previously discussed. By creating a feasibility matrix, the project team can assess each objective's feasibility and prioritize which ones to pursue. For example, the team could analyze the cost, resources required, and potential risks associated with each goal to determine their feasibility.

Additionally, the feasibility matrix can help the team identify any conflicting objectives or dependencies among goals, enabling them to make informed decisions about which goals to pursue and in what order. Overall, the feasibility matrix can help ensure that the project team is focusing on achievable objectives that will deliver the desired results for the project and align with the company's goals.

> **Important note**
>
> An important note to highlight about using the feasibility matrix is that it should be viewed as a flexible tool rather than a strict framework. The feasibility matrix is designed to guide the decision-making process and help project teams assess the feasibility of their objectives.

The need for iteration with research

After completing the feasibility matrix, the project team should have a clear understanding of which goals and objectives are most feasible and should be pursued. However, it's important to recognize that even the most well-informed decisions are based on assumptions and, in some cases, incomplete information. Therefore, it's crucial to continuously validate and iterate on research throughout the project's life cycle to ensure that the team is on track and making progress toward the desired outcomes. In this section, we will explore why iteration is essential in research and how it can improve the overall success of a project.

Continuous validation is a best practice in research that involves regularly testing and validating assumptions throughout a project's life cycle. By continuously validating research, the project team can ensure that they are on the right track and identify potential issues or roadblocks early on. This approach can also help teams identify opportunities to improve the project's outcomes or pivot their strategy if necessary.

To effectively implement continuous validation, the research team should regularly review and analyze the project's data and feedback from stakeholders, which is in itself simply another version of UX analytics. This feedback can help the team identify areas that need improvement and guide them toward making informed decisions about the project's direction. It's also important to recognize that there will be times when your business stakeholder partners are forced to make a decision that halts current progress and forces the start of a new direction. Continuous validation ensures that you don't go too far in one direction, and makes changes such as this easier and more cost-effective.

In addition to reviewing feedback, the research team should also continuously test and validate their assumptions. This involves designing experiments, conducting additional analytics through user testing, and gathering data to confirm or negate hypotheses that have been formed to this point. By testing and validating assumptions, the team can refine their understanding of the problem they are trying to solve and improve their chances of success.

Simply put, continuous validation is a critical best practice in research that can help project teams achieve their objectives and improve their overall outcomes. By remaining open to feedback, regularly testing assumptions, and adapting their approach as necessary, teams can ensure that they are on the right track toward success.

If we get into the experiments mentioned, we're back to what we've already learned – and are reinforcing here – A/B testing, prototype testing, surveys and polls, and more customer interviews. The point is that we're working on the best possible information we have at the moment, and as we iteratively work through the problems and create possible solutions, we're changing the interface and environment in which our users work. We make a change based on our analytics findings, and we test to see whether we're headed in the right direction.

Similar to driving a car, you must make small readjustments throughout the project life cycle. Are you with me on this one? You don't get in the car, get on the road, and then hold the wheel absolutely still on a straight road. You're constantly making small micro-adjustments with the wheel, accelerator, and brakes to keep the car moving safely. Your consistent use of UX analytics is the same for your project – keeping your progress on point and safe, or making the necessary adjustments to course correct.

So, by using a combination of these and other types of experiments, research teams can continuously validate their assumptions and make informed decisions throughout the project's life cycle. The key is to remain open to feedback, be willing to adapt the project's direction if necessary, and maintain a focus on the project's overall objectives.

Summary

This chapter highlighted the significance of the North Star statement in guiding a project's objectives and overall direction. The North Star statement is a concise and aspirational description of a project's ultimate goal, and it serves as a critical reference point for decision-making and ensuring that the project remains focused on its intended outcomes. By establishing a clear and aspirational North Star statement and regularly testing and validating assumptions, project teams can increase their chances of success and achieve their desired outcomes.

We learned about the formula for writing one and how we pull the purpose section out to write our North Star statement, and then also how to iterate through, using constant feedback from new testing to continually fine-tune and validate our work against the reactions of our users to the changes we're making with the product.

Coming up next: change management and the importance of realizing that any endeavor such as this is about much more than just the systems; it's about the people and their needs as well.

6

Change Management: It's Not Just about the Systems, It's about the People

Change management is a crucial aspect of any project and involves introducing new tools, processes, or systems into an organization. Even when the proposed changes are intended to bring about significant benefits, the human tendency to resist change can make the adoption process difficult.

> **Note**
>
> In fact, research conducted by McKinsey shows that as many as 70% of organizational change initiatives fail to achieve their intended goals due to a lack of effective change management.

To succeed in introducing new tools, processes, or systems into an organization, there are several key skills that UX teams should possess. These include learning about the mental models of stakeholders and users and understanding how they may resist change, designing the best possible rollout plan to mitigate disruption and ensure a smooth transition, and creating excitement and positive buzz about the new way of working to encourage adoption and buy-in. By mastering these skills, you and your team can significantly increase the chances of a successful implementation and maximize the return on investment for the organization. And this holds true whether you're tasked with change management, as many small teams are, or your team is acting as a support partner to the change management team in place.

In this chapter, we're going to dig through the underlying challenges inherent to change management and we'll cover these topics:

- Understanding the hard truths
- Taking baby steps to finish strong
- Evangelizing the benefits

> **Note**
>
> It's important to also point out that change management is a huge topic and can't be covered fully in this book. What you'll learn is how and why change management should be coupled with your UX analytics efforts for the best possible outcomes. This won't be an exhaustive covering of the topic so please do seek out books and blogs on the topic if you want to get a deeper understanding.

Understanding the hard truths

Let's consider the use case of a new internal order management application that's been approved by the business to meet the needs of a large user base of customer service representatives. While the new system promises to deliver significant productivity gains and cost savings, the transition to the new tool is likely to cause some disruptions.

Specifically, users will need to adjust their workflows and learn new processes, which can result in a learning curve and decreased efficiency during the initial stages of adoption. If the application is to be used by people who've been on the job and doing things their way for 10 years or more, this will be a very difficult transition for most of them. This is a hard truth.

To mitigate these challenges, a properly staffed and well-trained change management team is important, but as stated earlier, it's quite possible that your organization doesn't have a change management team. In this case, it will be up to you and your team to know what to do and to lead the efforts that your analysis has proven to be most effective toward the goals established.

By providing users with the necessary support, training, and resources, change management can help to ease the transition and ensure that the benefits of the new system are realized as quickly and as smoothly as possible. Best practices for change management include preparing the workforce for the change well in advance of the release, creating training materials such as videos and documentation, and holding short, guided practice sessions with power users.

Additionally, it is important to communicate the benefits of the new system and address any concerns or resistance that may arise during the transition process. By taking these steps, change management can play a critical role in ensuring the success of a new application or system and maximizing the return on investment.

Change management and mental models – stakeholders

In *Chapter 4*, we introduced the mental models concept and showed the four main types of thinking styles: auditory, visual, kinesthetic, and read/write. And, as written in the preceding paragraph, "*preparing the workforce for the change well in advance of the release*" means that we want to do our best to ensure that the people with all those different mental models can effectively learn what's needed quickly and with a level of depth that makes them skilled in the least amount of time.

By recognizing the four main thinking styles, teams involved in change management can tailor their communication and training strategies to meet the unique needs of each group. And you must count both stakeholders and users of the newer application as part of the change management sphere. The stakeholders have to understand all of the implications of incorporating the new system into the company, so you and your team have to include them as part of the efforts. And to do that most effectively, you need to know their thinking style.

For example, some stakeholders may respond better to visual aids such as diagrams and flowcharts, while others may prefer hands-on practice and role-playing exercises. By providing a variety of training materials and methods, teams involved in change management can help ensure that all stakeholders will effectively learn about the new processes and tools needed for successful implementation and how best to work with you to create the change management steps that will work best for the downstream organization, which includes your users and individual contributors.

Change management and mental models – individual contributors and users

To prepare your users for the transition to the new system, you should provide training materials that cater to each of these thinking styles, which is the same as for the stakeholders, with the exception that where there are 1 or 2 stakeholders, there could be anywhere from 10 to 1,000 users. Because of the scale here, it will be pretty much impossible to know which people fit into which category, so you have to prepare for all of them.

For auditory learners, we can create instructional videos and conduct live training sessions with a coach/instructor. Visual learners will have access to diagrams and flowcharts that explain the new system's processes. Kinesthetic learners can participate in hands-on training activities that allow them to practice using the system in a simulated environment. Read/write learners will have access to written instructions and manuals that explain the new system's functionality in detail, with quizzes that let them test their understanding as they progress through the material.

By tailoring our training materials to accommodate different thinking styles, we ensure that all users can learn the new system quickly and thoroughly. This approach will help to minimize disruptions and maximize productivity as we transition to the new way of working.

However, simply providing training materials isn't enough. It's important to also create a culture of learning and growth within the organization to support ongoing skill development and knowledge sharing. This can be achieved by encouraging peer-to-peer learning, providing opportunities for feedback and collaboration, and recognizing and rewarding individuals who demonstrate a commitment to learning and growth. By fostering a culture of continuous learning, you and your team can help ensure that stakeholders remain engaged and invested in the success of the new system long after its initial implementation.

Work this backward and you realize that because of the UX analytics you've done, and the interviews and other research work, you have a good picture of the personas and most likely have a very good feel for the user base and how they think.

Taking baby steps to finish strong

As I think you're starting to understand, change management can be a daunting task, particularly when it involves introducing new tools, processes, or systems into an organization that has done work in the same way for many years. In such cases, it's easy to get caught up in the excitement of the potential benefits and want to implement the changes as quickly as possible. However, taking a big-bang approach to change management can often lead to unexpected challenges and setbacks, leaving stakeholders feeling overwhelmed and frustrated.

That's where the concept of taking baby steps comes in. By breaking the change management process into smaller, more manageable steps, you can minimize disruption, reduce resistance, and set your team up for success. So, let's explore how to design and execute an effective baby-steps rollout plan to ensure that your change management initiative finishes strong.

When a project is approached in smaller increments, it's less disruptive and easier for team members to adapt to changes, which ultimately results in a more successful outcome. The goal is to design and execute a baby-steps rollout plan that will facilitate an effective change management process.

When projects are worked in an Agile fashion, this is much simpler to achieve because Agile is based on small steps that are tested, realigned, worked, and so on. A good change management plan can be designed to work in the same way.

Here's a five-point process for achieving a baby-steps approach:

1. Determine the goals and scope of the change management project and break them down into smaller, more manageable steps. This will help you to focus on what's most important and to identify the tasks that can be tackled in small increments.

2. Create a phased approach to the project, where each phase is broken down into smaller, incremental steps. This approach will help you to focus on specific tasks and milestones and to track progress along the way. If your company uses Agile development, this fits in neatly with that management style.

3. Prioritize the tasks within each phase based on their importance and level of difficulty. Start with the easiest and most important tasks first, and then move on to more complex ones as you progress.

4. Communicate the phased approach and priorities to all stakeholders and involve them in the process. Encourage feedback and collaboration and address any concerns or resistance that may arise.

5. Monitor progress throughout the process and be prepared to adjust the approach as needed. Regularly review and adjust the priorities, tasks, and timelines based on feedback and progress made, and continue to communicate and involve stakeholders throughout the process.

Now that we understand the power of working in small steps and phases to ensure that we get where we intend with the least possible amount of rework and wasted hours and effort, let's move on to how we turn all the work into a plan of action.

Designing a solid rollout plan

Your rollout plan will be a collaborative effort with product and engineering, specifically, someone from your training department (if you have one) so that the proper training materials can be designed and produced, and sometimes marketing if your product is an external one. It will require a well-planned and well-executed approach that considers the unique needs and challenges of the organization. To achieve this, it's important to identify the project goals and scope and to break them down into smaller, more manageable steps that can be tackled in a phased approach. Coordinating with the HR department is also a good idea, as that will enable them to be fully aware of the coming changes and create their own material for employees and onboarding.

Developing a phased approach allows you to focus on specific tasks and milestones and to track progress along the way. It's important to prioritize tasks based on their level of importance and difficulty and to start with the easiest and most important tasks first. This approach helps to build momentum and confidence among team members, which can make the later, more complex tasks easier to manage.

Effective communication and involvement of stakeholders throughout the process is also key to success. Encourage feedback and collaboration, address any concerns or resistance that may arise, and be prepared to adjust the approach as needed. By following these steps, you can develop an effective baby-steps rollout plan that will help you to achieve your change management goals while minimizing disruption and resistance.

Let's work with an internal rollout as an example, and we'll use the order management system for customer service reps illustrated earlier. You need the lead UXer, the product manager, an engineering lead, and the training department manager or lead (if you have one) to be the foundation of the team. There will be times when you need to bring in other SMEs to help refine plans, but this is the team that will be responsible for the rollout plan definition and execution.

So then, the first step in developing an effective internal rollout plan is to get the work done in scoping and defining the project, as you completed when creating the North Star statement and all the work surrounding that. This needs to include the specific features and functionality of the order management system that will be rolled out to customer service reps, as well as the timeline and budget for the project.

Once the goals and scope of the internal rollout have been defined, the next step is to establish a phased approach for the rollout. This includes breaking the project down into smaller, more manageable steps that can be tackled in a logical order and assigning specific tasks and responsibilities to each member of the team. If, for instance, there are hardware needs that must be met before any other work can progress, this is most likely a set of tasks that your engineering team will carry out, ensuring that the team is informed of progress or blockers that might slow down other aspects of the project.

> **Using a RACI chart**
>
> **RACI**, if you're unfamiliar with it, stands for **Responsible, Accountable, Consulted**, and **Informed**. This is a matrix type of chart that names the project and all the people involved in the tasks to completion. In our preceding example, while the engineering team is carrying out their tasks of getting and installing hardware, they're keeping other members of the team informed.

As mentioned earlier, the lead UXer, product manager, engineering lead, and training department manager/lead will form the foundation of the internal rollout team. However, it's important to assign specific roles and responsibilities to each team member to ensure that everyone is clear on what they need to do to contribute to the success of the project. If you decide to use the RACI form, each member will be listed in rows, and then marked as **R**, **A**, **C**, or **I** accordingly.

Effective communication is critical to the success of any internal rollout plan. It's important to develop a comprehensive communication plan that includes regular updates to stakeholders, as well as training and support for customer service reps who will be using the new order management system. This will tie into your RACI if you have your stakeholders listed, and the communication to them can be via emails, Teams, Slack, or any other communication channel you use in your company.

The final step is to constantly monitor and adjust the rollout plan as needed. This includes tracking **key performance indicators** (**KPIs**) to measure the success of the rollout, as well as soliciting feedback from stakeholders and making any necessary adjustments to the plan based on this feedback. By following these steps, you can develop an effective internal rollout plan that will help you achieve your goals while minimizing disruption and resistance.

Here's an example of what a RACI chart might look like for this project. You'll see other examples and versions online that have color coding, so design and use what suits you and your team's needs best. Quite often, a project manager will be involved, and they might take care of this. However, different companies run their teams in many ways, so it's always best for you as the UX team to understand this and be able to lead the efforts by creating one and managing it for the team. You did all the analytics that got the project started, so be involved with this as well to show how UX leads.

Task	Responsible (R)	Accountable (A)	Consulted (C)	Informed (I)
Define user requirements	UX lead	Product manager	Engineer	Change management team
Create wireframes and prototypes	UX lead		Engineer	Change management team
Develop and test code		Engineer	UX lead, Product manager	Change management team
Develop project timeline and budget	Product manager		Engineer, UX lead	Change management team
Develop and deliver training program	Change management team	Product manager	UX lead, Engineer	
Conduct user testing and feedback	UX lead		Engineer, Product manager	Change management team
Monitor project progress and milestones	Project manager	Product manager	Engineer, UX lead	Change management team

Figure 6.1 – RACI example for the project

In this example, the responsibilities are divided among the team members as follows:

- The UX lead is responsible for defining user requirements, creating wireframes and prototypes, and conducting user testing and feedback.

- The engineer is responsible for developing and testing the code and is consulted by the UX lead and product manager.

- The product manager is accountable for developing the project timeline and budget and is consulted by the engineer and UX lead.

- The change management team is responsible for developing and delivering the training program and is consulted by the UX lead, engineer, and product manager. They're also informed of the project's progress and milestones by the project manager.

My biggest takeaway from the use of a RACI chart is how it keeps everyone accountable. It doesn't matter which responsibilities fall to you; you are accountable in some way to the project. Even if you're on a team that's simply informed of certain tasks and actions, you're accountable for reading what comes to you, understanding that, and – most importantly – speaking up if you see a challenge in the way the project is proceeding.

You can see that a RACI chart helps to clarify responsibilities and ensure that everyone is on the same page. This is especially important in complex projects where there are many moving parts and different stakeholders involved. In addition, it can also help to identify any potential gaps in the project, such as areas where there may be a lack of accountability or unclear responsibilities.

By clearly defining roles and responsibilities, team members are more likely to work together and share information. This can lead to better decision-making and problem-solving, as team members are able to draw on each other's expertise and experience, which, in turn, can help to drive project success.

Evangelizing the benefits

When it comes to software application projects, the success of the project isn't just about the technology being used or the features being developed. It's also about the people involved in the project and how they are able to adapt and work together toward a common goal. But change management is bigger even than just that. Proper change management encompasses the entire organization: C-suites, management, stakeholders, teams, and individual contributors.

This is where project evangelism becomes critical. By evangelizing the need for the changes to increase efficiencies and achieve company goals, and by using management tools such as the RACI matrix, your team and project managers can help to ensure that all team members understand their roles and responsibilities and are committed to the success of the project. That's number one, of course. Next and equally as important – maybe even more so – is the need to evangelize down to the user level. You must show the users of the new application or system how effective they'll become by adopting it. Since people are by nature resistant to change, your work will represent a change to them. You interviewed and researched and found the pain points. You collaborated with product and engineering teams to create new workflows and features to eliminate those pain points. But remember that even though you were told, or you discovered through the tools, that there were problems in the current application, people got used to it.

They may have created workarounds that, while not efficient in actuality, seemed more efficient to them. I've seen production managers in manufacturing facilities create spreadsheets out of frustration with how difficult software systems were to learn and use. It took one person five months to perfect his spreadsheet, but to him, that was less painful than using the new system the company bought and rolled out without much thought to UX.

The users need to understand how the new application will make their work easier, faster, or more efficient. One way to do this is to highlight specific pain points that the current system has and how the new application will solve those problems. For example, if the current system is slow and cumbersome, emphasize how the new application will be faster and more intuitive.

Another effective approach is to involve users in the development process from the beginning. This can be done through focus groups, surveys, or other feedback mechanisms. By involving users in the process, they become invested in the success of the project and feel like their input is valued. This can lead to a sense of ownership and buy-in for the new application.

Finally, it's important to communicate clearly and consistently about the changes that are happening. This includes explaining why the change is necessary, what the benefits are, and what the timeline for implementation will be. By being transparent and upfront about the changes, users are more likely to feel informed and prepared, which can make the transition smoother and more successful.

Understanding who else the change will affect

While the users of a new system are our key group because this is a UX project, they aren't the only people who'll be affected by the change. Developers will have to adjust to a new environment, and we must be careful to ensure that they're able to adjust to these changes without disrupting the development process. Engineers will also be affected if any of the backend systems are changed and the architecture is altered in ways that cause major system infrastructure retooling to take place.

Helping developers adjust

Developers need to understand why changes are being made and how the changes will impact their work. This means providing regular updates and information about the changes, as well as giving developers the opportunity to ask questions and provide feedback.

In addition to clear communication, it's important to involve developers in the change management process as early as possible. This means including them in the planning and design phases of the project and giving them the opportunity to provide input on how the changes will impact their work. By involving developers in the process, they will feel more invested in the changes and more willing to adjust their work processes to accommodate them.

Because they tend to focus more on the day-to-day tasks of interface coding, testing, and maintaining the software, their work will be closer to the end user. They may work on individual components of a larger system, or they may be responsible for the development of entire applications or products. Either way, developers in most companies that have both development and engineering teams are close to product managers and the UX team and will be affected by all the changes that will make the products easier on end users.

Helping engineers adjust

Software engineers tend to focus on the larger-scale design and architecture of software systems, with a focus on ensuring that they're efficient, reliable, and scalable. They often have a background in computer science or software engineering and may specialize in a particular area such as data engineering, infrastructure engineering, or security engineering. They use a range of tools and methodologies to create software systems, including design patterns, UML diagrams, and software engineering processes such as Agile or Waterfall.

The engineering team plays a critical role in the design, development, and implementation of any new software application, so it's important to get their buy-in early on in the process. They must understand and support the goals and scope, the timeline, and milestones so that they can plan the work that everything will be built upon, and the need for any new technologies or frameworks that will be used in development.

They need to be aware of how the new software application will impact existing systems as well. They need to understand how the new application will integrate with existing systems and what changes will be needed to ensure compatibility. All of this will cause disruption to current-state application foundations if the newer software requires a rebuild. By addressing their needs and concerns in a change management effort, you ensure that they can support the new software application effectively.

So, you see now that the need to evangelize the benefits must be tailored to a group larger and wider than just the end users of the new application. Change affects everyone who will touch the work in some way, shape, or form. By including development and engineering teams in that scope of people to serve through thoughtful change management plans, you can increase the chances of successful, on-time integration, while reducing the need for costly reworks or ambiguous and overlapping tasks.

Summary

Change management is one of the cornerstones of successful application redesign or introduction projects. It's a wide-scope effort that must include management down to end users. You and your team did the work to prove the need, through the pains and challenges of users. When you combined that with a workable path forward, the project was adopted.

Now, you must take everything that will change into account so that you can manage expectations and provide the proper information and training to those who will be affected. You must articulate the problems in ways that make the picture crystal-cear for everyone, and then show the solution in a manner that allows everyone to understand the reason for the change and how it will benefit them in being more efficient with their tasks.

If you always remember that people are creatures of habit and resistant to change, you'll also never forget to include a proper change management plan in your work to deliver the best new product possible.

Next up, in *Chapter 7*, we're going to get into the internationally accepted practice of design thinking and show you why it's the absolute best tool for collaborative problem-solving.

Part 3: Research

This part of the book delves into the critical role of research in the field of UX, equipping you with the knowledge and techniques necessary to conduct effective and insightful research practices. At the core of UX lies design thinking, a visual and collaborative practice that drives the discipline forward. This part begins by guiding you through step-by-step processes and best practices to become respected facilitators of design thinking. Establishing a core team is the initial step, followed by conducting various levels of interviews to solidify management's expectations and align them with input from managers and individual contributors.

To gain a comprehensive understanding of the problems at hand, it becomes essential to venture into the market and capture the broad sentiment surrounding those problems. Surveys provide an efficient means to achieve this, allowing for quick data collection and analysis beyond what in-person interviews can offer. The chapter on surveys explores their benefits and strategies for effective implementation.

UX metrics play a vital role in measuring the impact of user experiences. That chapter delves into the two types of metrics: qualitative and quantitative. Qualitative metrics focus on feelings and sentiment, providing insights such as preferences or pain points expressed by users. By quantifying and analyzing these qualitative metrics, practitioners can identify areas of focus and make data-driven decisions.

Understanding data is fundamental to any analytics practice, and this understanding must encompass multiple perspectives. The UX analytics practitioner needs to develop an intimate familiarity with all measurable aspects of the business to support management goals effectively. This includes comprehending the taxonomy and hierarchies that define data relationships, enabling effective filtering and representation of information.

As for research tools, a wide range of options exists in the market, each catering to different needs and preferences. This part provides you with an overview of some of the best tools available, including links to their respective websites. You are encouraged to conduct your due diligence to determine the best fit for your specific requirements, company, and the nature of your work.

This part has the following chapters:

- *Chapter 7, Design Thinking – Understanding the Core Problem-Solving Tool in a UX Practice*
- *Chapter 8, Interviews and Personas – Best Practices for Quick and Efficient User Understanding*
- *Chapter 9, Surveys – A High-Level, Broad-Stroke Tool for Quick Direction*
- *Chapter 10, UX Metrics – The Story of Operational Measurement*
- *Chapter 11, Live by Your Data, Die by Your Data*
- *Chapter 12, The Tools It Takes to Gather and Consolidate Information*

7

Design Thinking: Understanding the Core Problem-Solving Tool in a UX Practice

The heart of a design thinking discovery session is the cross-functional team, and you must know how and why to put this team together. The right team will bring subject expertise from different aspects of the company, and it's those different viewpoints that come up with the best, most well-rounded ideas as the foundation for analysis.

In this chapter, we're going to explore the team concept in depth and get an understanding of why it's the key element in your design thinking practice. You'll learn about the processes, the team, and how to master the art of discovery session facilitation. Each part of the process is essential to successful outcomes.

And as this chapter is in the research part of the book, we'll get a good start and a glimpse of the coming chapters, each covering specific, detailed aspects of solid research work and how it propels everything you do in UX analytics.

In this chapter, we will be covering the following topics:

- Exploring the core processes of design thinking
- The core team
- Discovery session facilitation – learning and mastering the fundamentals

Exploring the core processes of design thinking

Design thinking is often misunderstood because of the word *design* in the title. Some people think that it only deals with the visual design of an application, the interface. And they feel that they have no expertise or sometimes even no interest in that.

But that's far from the real discipline. Yes, there's a part further on where we may start to wireframe, depending on the ask, but it's only a supporting part of the overall practice of design thinking. This is just a nuance, but to me, Design Thinking should always be capitalized because it's a verified and widely used professional practice and system for solving problems.

Research is the foundational element of Design Thinking because we're working toward, first, an understanding of a current state situation. We diverge to get as much information as possible from our team and the people we interview, then converge down to a focus set based on the answers and ideas in our categorized groups.

Nobody does this alone. Design Thinking is a team sport! Everyone participates in a way that brings out all the best information about processes, systems, people, and data. A core team is your hand-picked selection of experts that will be with you for the duration of the project, so it's important to ensure that they commit to the time it takes to get the work accomplished.

If you've got a fresh team, people who've never done this before, you start with an empathy exercise to get them out of their heads and into a place where they can put themselves into the shoes and work challenges of a persona. We do this by creating a scenario where they must play a role. This is one of my favorites – please use it to start your journey if you'd like.

Setting the scenario

There's a download for this that you can print out from the book's GitHub repository (`https://github.com/PacktPublishing/UX-Analytics-Best-Practices-Tips-and-Techniques-`) but here's a snapshot of it:

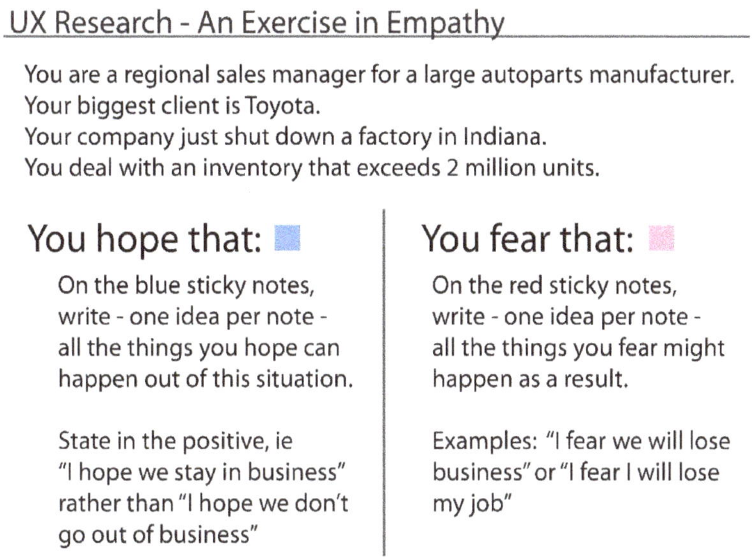

UX Research - An Exercise in Empathy

You are a regional sales manager for a large autoparts manufacturer.
Your biggest client is Toyota.
Your company just shut down a factory in Indiana.
You deal with an inventory that exceeds 2 million units.

You hope that:

On the blue sticky notes,
write - one idea per note -
all the things you hope can
happen out of this situation.

State in the positive, ie
"I hope we stay in business"
rather than "I hope we don't
go out of business"

You fear that:

On the red sticky notes,
write - one idea per note -
all the things you fear might
happen as a result.

Examples: "I fear we will lose
business" or "I fear I will lose
my job"

Figure 7.1 – Empathy exercise

Part 1 – You print this out and tape it up on the wall. Next, you have someone from your team read it out loud to the group. Then you read the directions:

- On blue sticky notes, you write things that you hope will happen.

- Don't phrase anything in the negative. In other words, don't write "I hope we don't have to fire people." Instead, write "I hope we can keep all our staff."

- On the red sticky notes, write what you fear might happen.

- Put all the blues on the left and all the reds on the right. If you make a mistake and write a fear on a blue, write it over.

- Write your idea even if someone else wrote it or similar. This helps with the next part of the exercise.

- You'll have three minutes. The time starts now.

- Give a time check at two minutes.

- Give a time check at one minute.

- Give a time check at 30 seconds.

- Count down the final 10.

- If anyone is writing, have them finish the thought and put it on the wall.

This first part is individual work. They should all be writing their own ideas and not discussing what they're writing. They're each, in their own words, expressing what they'd be feeling if they were that sales manager. The collaboration will come next.

Part 2 – Now, everyone will start to group similar themes. These are usually finance, workforce, plant closure, inventory, and a few others. All the sticky notes with the same idea or sentiment should be grouped together.

Have them form columns as in the following illustration:

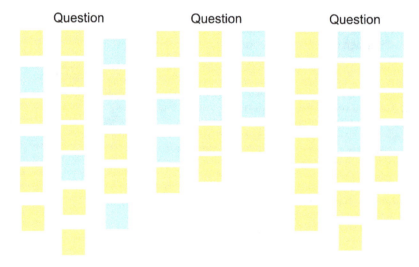

Figure 7.2 – Categorization exercise

As the facilitator, you're ensuring that everyone is participating, starting to read some of the stickies out loud, and keeping the energy high.

This exercise serves two very important purposes:

- It teaches them about empathy so that when you start interviews, they have the concept firmly planted and can take notes and ask clarifying questions from a fresh perspective
- It teaches them the processes of collaboration and iteration that will be used throughout the session or sessions

Okay, that's a good start on Design Thinking. Next, let's dive into the team and the facilitation processes now.

Assembling your core team of **subject matter experts** (**SMEs**), as mentioned in the introduction, this team is a collection of people whose knowledge and experience will benefit the creation of the UX analytics strategy. You'll start to put this team together as soon as the project outline is set by the business owners or stakeholders handing it off to you.

With some company types, it can be possible to have this team set and ready to go for any project. With others that may have a more varied workload and a wider array of product types, this team can be assembled based on the need. This is why it's important to you, as the UX analytics SME, to get to know all the key players in your company. You need the company's trust so that you can make a phone call or send an email and get the time you need from the SMEs for your team.

Remember that analytics is about data, and data is the new gold. It's the currency by which companies will thrive or die. Because you're going to have SMEs on your core team, you're going to have access to the people who can support the capture and use of data – outside of the tools you'll use to do automated analytics.

As well as providing access to data at their levels, they can open the doors to do the research you need to complete with users. And these users will have different levels of data access and different jobs to do in supporting the enterprise.

Let's look now at the team members and their roles.

The team facilitator

The facilitator is the person whose job it is to ensure that everything is set and ready to go for the sessions you have. Whether virtual on a whiteboard program or in person, they'll be in charge of all materials, timekeeping, setting the schedule, and getting the team members scheduled and set. There will often be times when this person will also act as the primary interviewer, a role we'll discuss in a bit.

The facilitator will provide sticky notes, Sharpies (never pens or pencils), sticky dots for voting, and wireframing templates if that will be part of the work. If your team will be traveling to a customer site, box this up and send it ahead of time so that it's there when you arrive. Always do this and don't rely on a customer or other site from your company to provide these things. You need to control this so that there are no mistakes.

The facilitator will also be running all exercises and doing their best to make sure that everyone on the team participates. The facilitator is the impartial participant and leader of the sessions. They must be able to control the room and situations that may arise.

In the following section, we'll get deeper into the role of facilitator and provide some real-world examples of facilitation in practice.

The primary interviewer

This role has a very important job to do. They must run interviews when you have them and participate in all other work when interviews aren't scheduled. The primary interviewer runs the interview sessions and, as the first contact with the interviewee, must establish a rapport with them so that they give honest answers to the questions, which will then become part of your analytics data.

I always suggest that when working in person at a customer facility, this person has the customer stakeholder or another manager walk them around and introduce them to the people who will be interviewed. You say hello to them and let them know that they'll be coming in to see you later. This way, they don't walk into a room cold. You're now a familiar face, and if your travel team consists of three or four people, which is reasonable for larger projects, they won't know any of the rest of the team.

They may very well know a few of the customer core team, and they may not. So doing this simple step allows you to ensure that they walk into the room with more confidence.

The primary interviewer will introduce the table to the interviewee, and then the primary, and only the primary interviewer will be asking the questions from the interview blueprint to start. The primary interviewer never takes notes – that's the job of the others on the team, and we'll see some details on them in the next section.

The primary interviewer orchestrates the entire interview and must establish a rapport with the interviewee so that your team gets the best, most honest, and most accurate answers. You must assure them that anything they say will be in confidence and their name will not be revealed to anyone. They have to understand this, and it's a key best practice of UX research – psychological safety. *Chapter 8* will dive into this in greater detail.

The primary interviewer, once they've gotten the interviewee comfortable, will sit face to face with the interviewee, as in the following figure:

Figure 7.3 – Seating arrangement when interviewing

This way, the focus can remain on these two people. Never have a table or anything else between you. Keep a warm, open smile on your face, and only look at the blueprint to ensure you're following the correct order of questions and how they should be phrased. This may seem daunting at first, but after a few, you'll gain a rhythm, and the interviews will start to flow.

Once the primary interviewer completes the questions, they will then direct the others – one at a time – to ask any clarifying questions they may have, which can go like this:

Primary looking to their left at the table:

"Mary, do you have any clarifying questions for Bill?"

"Yes, thanks Jeff. Bill, when Jeff asked you about your challenges, you said that data access is tough. Can you tell us a bit more about that?"

Now, Bill will answer Mary. She may have another question, or she'll indicate that she's good. Then, the primary calls on the next person, and the next, and so on, until they've hit everyone taking notes. This serves to allow the team and the interviewee to slowly get to know each other. Never, at any time, should core team members interrupt the flow with questions, either during the primary phase or otherwise. Keeping soft control is the best way to build momentum and allow for the relationships between the team and the interviewees to flourish.

Okay, so now the interview is complete, and the interviewee leaves. The facilitator will take over again, and those duties will be discussed in that section. Let's talk about the others in the core team now.

Attributes of successful interviews

So, what are the core team members responsible for during interviews? There are a few things, and they're very simple:

- **Listening intently**: When listening to an interviewer and interviewee interacting, it's important to focus all attention on that interaction. Look at the interviewer, with glances to the interviewee to see how they're reacting to being asked the questions.

  Be cautious about letting your mind go to solving the problems being stated. Often, we hear a problem or a statement and start thinking about how to fix it or what's next. Keep focused on the conversation.

- **Watching body language**: People give clues other than verbal ones when being interviewed or in conversation. Watch for facial expressions, eye contact, and body postures. If the interviewee is sitting with their hands in their lap or on their knees and then suddenly switches to crossed arms when asked a certain question, it could mean that they're being guarded and possibly not being honest.

- **Taking notes**: The team will be taking notes in the blueprint as the primary interviewer conducts the first round.

Keep the notes short and concise. Refrain from writing sentences or being verbose because these are harder to decipher and transcribe in later steps of the process.

Highlight the areas you want to ask clarifying questions on later so that you don't have to try to remember them. Just a simple star or check mark is fine – anything to help you find it quickly when it's your turn to ask a question.

- **Asking clarifying questions when directed by the primary interviewer**: When the primary interviewer calls on you, do as in the example that we saw earlier. Mention the question that was asked, what you wrote down for their answer, and then ask them your clarifying question. The key point here is to try as best you can to not be scribbling notes while the person is talking to you. Try to stay focused on them, and then quickly jot down a few notes when your turn is over.

When others are asking their clarifying questions, take notes on those as well. This helps them and the entire team with capturing the most accurate information possible from the interview.

The role of the core team members during iteration

The interviews are over and each team member has a small stack of blueprints filled with notes for each interviewee. Everybody's will look different except for the fact that they're all contained in the blueprints.

Iteration is the process of divergence and convergence. Look at the following figure:

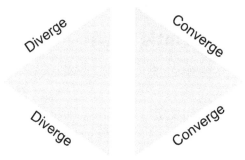

Figure 7.4 – Diverge and converge sequence

The start is on the left, with nothing. As we do the work, our board begins to fill with sticky notes until we have all the notes written. Each person on the core team will transfer their notes for each question onto sticky notes, and all those will be put on the board. You'll have a section for each of the questions asked. At the end of the diverge phase, your board could look like what we showed in *Figure 7.2*.

Note that there are a different number of stickies in each section. This will happen naturally because each note-taker will write a different set of notes for each question. Note also that there are two colors of stickies. This is one way of showing that more than one person said the same or a similar thing. It's especially helpful when we're focused on the challenges and pain points.

> **Important**
>
> Each sticky note must have only one idea on it. This is extremely important for the converge phase. Any sticky that has more than one idea must be rewritten and split into separate stickies.

If we look at our diverge/converge illustration again, as seen in *Figure 7.4*, we're now starting on the converge phase.

You should have many sticky notes on the wall, and quite a few of them should be duplicates. That's the way you need it to be. If six different people said roughly the same thing, that's an indication that something needs attention: that's a point of focus.

It won't be so much that all core team members wrote the same things for one person; it will be that several people interviewed said the same thing. I must caution you and the team here. Don't rush this part of the process. It does take time, but it's the best way to get the job done right. Knowing where there's a focus and seeing where there may be some outliers is critical.

A real-world example of design thinking

At a hospital in Canada, my team and I first interviewed stakeholders to get a feel of what the hospital felt needed to be addressed. We then interviewed five ward managers, each responsible for staff, equipment, materials, patients, schedules, and so on. They were in charge of everything within that department. Think of the cancer ward, neonatal, or cardiac care.

When we went through the diverge/converge stage, a definite pattern began to arise. One ward manager worked very differently than the others. When all the work was complete, there was a big grouping of stickies on the left of the wall and then one very compact grouping on the right. You would normally see those as outliers, but in this case, it was evident to us and the stakeholders that this manager worked differently.

The catch was that he was also the most efficient manager in the hospital, and his success couldn't be denied. Because of these findings, which had never been brought to light in the hospital, the stakeholders were forced to change their position on the work to be done. We regrouped, took a bit of a pivot, and rewrote the plan to include more training for the others so that they too could work like that manager.

The cost to the hospital decreased considerably because there was now less work for our team to do, while overall efficiency was raised because of the streamlined workflow and ways of using the software.

So, now the team has a focused set of problems to be solved, or a definite direction to move in, based on the interviews and the Design Thinking work. Maybe you've uncovered five key areas of improvement to tackle, so what do you do now?

Consolidating, interpreting, and reporting

Because this part of the research doesn't deal with any metrics gathered from tools that measure clicks or time spent on page x, y, or z, you should wrap this up and report your findings. The team is finished with direct interactive work and everyone is back at their home or office desks. Now everyone has a role in completing the analytics task.

And now that we're embracing virtual work, the best thing to do is create a whiteboard for the task. Use **Mural**, **Miro**, or any number of other available tools – and it's probable that your company already uses one, so fire it up – and create a new board. If the facilitator handles this work, which is advisable, they can invite all the core team members to it.

The goal here is to consolidate all stickies into their respective categories, that is the questions asked, minus people's names.

Let's look at the user interview blueprint and talk through that:

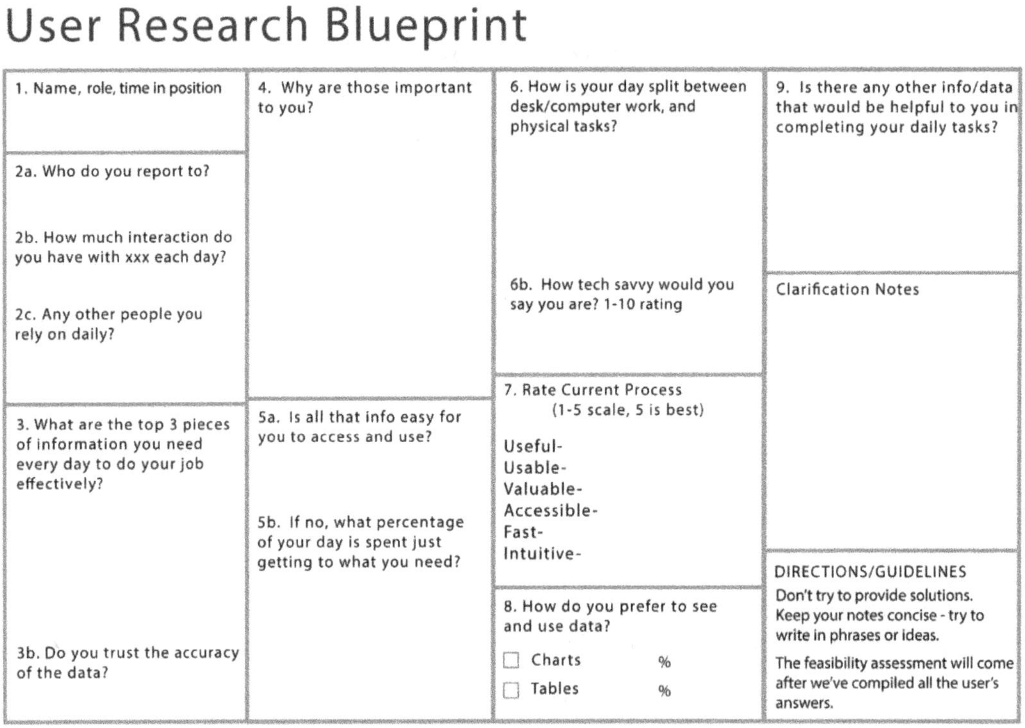

User Research Blueprint

1. Name, role, time in position	4. Why are those important to you?	6. How is your day split between desk/computer work, and physical tasks?	9. Is there any other info/data that would be helpful to you in completing your daily tasks?
2a. Who do you report to? 2b. How much interaction do you have with xxx each day? 2c. Any other people you rely on daily?		6b. How tech savvy would you say you are? 1-10 rating	Clarification Notes
3. What are the top 3 pieces of information you need every day to do your job effectively?	5a. Is all that info easy for you to access and use? 5b. If no, what percentage of your day is spent just getting to what you need?	7. Rate Current Process (1-5 scale, 5 is best) Useful- Usable- Valuable- Accessible- Fast- Intuitive-	
3b. Do you trust the accuracy of the data?		8. How do you prefer to see and use data? ☐ Charts % ☐ Tables %	DIRECTIONS/GUIDELINES Don't try to provide solutions. Keep your notes concise - try to write in phrases or ideas. The feasibility assessment will come after we've compiled all the user's answers.

Figure 7.5 – The user research blueprint

You can download this blueprint from: `https://github.com/PacktPublishing/UX-Analytics-Best-Practices-Tips-and-Techniques-`.

We'll cover this in depth in *Chapter 11*, so for now, just take a look through it and familiarize yourself with it. When we do our sticky notes consolidation, we don't use the person's name, just their role. Questions 1 and 2 are background info that can help us later but are not necessary for our direction unless we start to notice a pattern where this could help us get through a challenge faster.

The questions we'll be working with are 3 to 9. We'll take averages for 6b, 7, and 8 and fill in with answers from 9 to see whether there's a definite silo or gap we should be addressing. I'll emphasize again that this takes time – there are no shortcuts. The teams that get this right spend less time chasing things that don't matter and help their enterprises get ROI faster than teams that try to cut corners and speed through this process.

Mastering the fundamentals of discovery session facilitation

The fundamentals of session facilitation are simple: be organized, be direct yet respectful, be diligent about avoiding biases, and always be watching. A great facilitator is one who can practice empathy, know when a pivot or even just a break is needed, and can articulate processes, directions, and pace to a wide variety of personality types. Being non-threatening and persuasive is the key that will unlock cooperation and collaboration with the majority of people you encounter.

You're the leader, the coach, the arbitrator, and at times the voice of reason. You must stay impartial even if you've got strong opinions. Always work on team members' ideas and help to solidify or reject them based on what the data is telling you.

This part and the first phase is only research. There will be a time when your team will start solutioning, but this isn't the time for that. Put any of those ideas in a parking lot and use them when the right time comes.

Tone matters

The tone of your voice carries many subtle cues to those around you, especially when you're facilitating discovery sessions. Having the capacity to remain calm and alert during sessions will not only bring your team to their maximum performance, but it will also bring the project to its best conclusion.

There are people who may trigger you. It's a natural human response to the ways in which people speak and use their own tone. Some find it hard to express their frustration or stronger feelings without sounding like they're angry or irritated. While the rules of engagement at the beginning will state that the sessions are a safe space, they're also a time of professional respect. This doesn't mean that everyone, every time, will be able to control themselves.

This is because if the right team is chosen, your core team members will all have passion for what they do. They may fight especially hard to win a point when no one else agrees. You, as the facilitator, must be able to take that energy with grace and turn it into a positive outcome for the team. You can rephrase, ask that person – very calmly – to talk a bit more about why they feel the way they do, and you can also table certain discussions for later, when perhaps a smaller group of people or a stakeholder can discuss and solve the issue.

On the other side of this is the enthusiasm that comes when the whole team agrees on a direction. People – me included – can get very animated when talking up a point or rejoicing when a difficult situation is resolved. The point here is that you want to keep the mood high and positive as much as possible. If the team members are all friends who've been working together for years, the overall mood will most likely be lighter – with jokes and friendly banter.

Being a facilitator isn't always easy, but the rewards for a job well done are worth the effort. If you can master the tone and set the mood or change it up if needed, you'll be a much sought-after member of the UX team and will get to work on many cool and unique projects.

Always coach and encourage your team members to stay concise, don't overthink things, keep working, and so on. When you see someone doing great work, call that out. Mention each team member by name. *"Awesome, John. We've got 10 minutes left and you've already written a lot of great notes. Great job." "Alice, I like the way you wrote that one. It really keeps the user in mind. Well done."*

Time-keeping for focus

Every part of your discovery session should be timed. From two-minute sticky note phases up to thirty-minute diverge/converge phases, each part of a session needs to be sharp to allow for the correct flow within the team.

For the shorter parts, start the timer, instruct the team to start, and watch them. See who the strong members are and who the weaker ones are. This works very well when we start a new session with an empathy exercise. You read the directions in that section, so there's no need to go over those again except to remind you to set your timer, call out time segments, and wrap it up when the time is up.

If you're working virtually, on a whiteboard such as Mural, there's a time at the top and a signal when the time is up. Those are handy as a visual reference, but you still want to call out milestones. When working in longer stretches, such as 30 minutes or longer, make a call every 10 minutes.

Summary

That was a long chapter, and there were many key takeaways for you. You learned that Design Thinking is a crucial part of any analytics project that aims to get to the root of the problems faced. By getting direction from stakeholders and then speaking to users, you get to the *why* of problem areas you may see when you use analytics tools to watch and gather metrics from user interaction in your applications.

You learned about tone of voice and how important that is to ensure that your team and your interviewees feel comfortable and valued for their participation. Sometimes there will be strong personalities on the team, and making sure they participate with respect for their peers will always provide the best results in the end.

And since UX is as much about visuals as it is anything else, you saw a few illustrations that will help you understand the concepts faster. We're all visual beings – some more so than others – but seeing and recognizing patterns is one of the strongest traits of human beings.

All this can and should be used to your advantage in understanding Design Thinking, no matter the role you might play on a team.

Next up is a closer look at the interview process and personas. We started looking at these in this chapter, and now we'll get to the finer points of each.

8

Interviews and Personas – Best Practices for Quick and Efficient User Understanding

Understanding the ask from business stakeholders and new goals while considering problems from the bottom up is critical to a successful analytics project outcome. Any project that proceeds merely from the top down is likely to fail because the challenges of the users aren't considered.

Likewise, just working toward fixing user problems without input from the business and understanding company objectives is just as risky. While we do need to address the pain points; we can't do it in a vacuum. We must take these problems to the business in a way that they see the impact and results of current situations and the increased risk of not fixing the problems.

This chapter will therefore break this subject down even further. You'll recall from *Chapter 7* that we started to outline and dig into the roles of the core team and their responsibilities. So now we're going to learn about the two main personas: the stakeholder and the user. When framing an analytics project for any new product, this is where you should start.

In this chapter, we will cover the following topics:

- Getting comfortable with an interview protocol
- Dissecting the stakeholder interview and persona
- Grasping the user interviews and personas
- The importance of interviewing one person at a time

Getting comfortable with an interview protocol

In each of the two interview templates we're going to cover, it's important to pay close attention to the way the questions are worded. While it's always a better idea to be conversational and not just read questions to the person being interviewed, you do want to stick as close as possible to the wording.

Let's take a look at a quick interview sample to give you an idea of this concept. We'll work on a user interview because you'll do many of these, as opposed to one or maybe two stakeholder interviews.

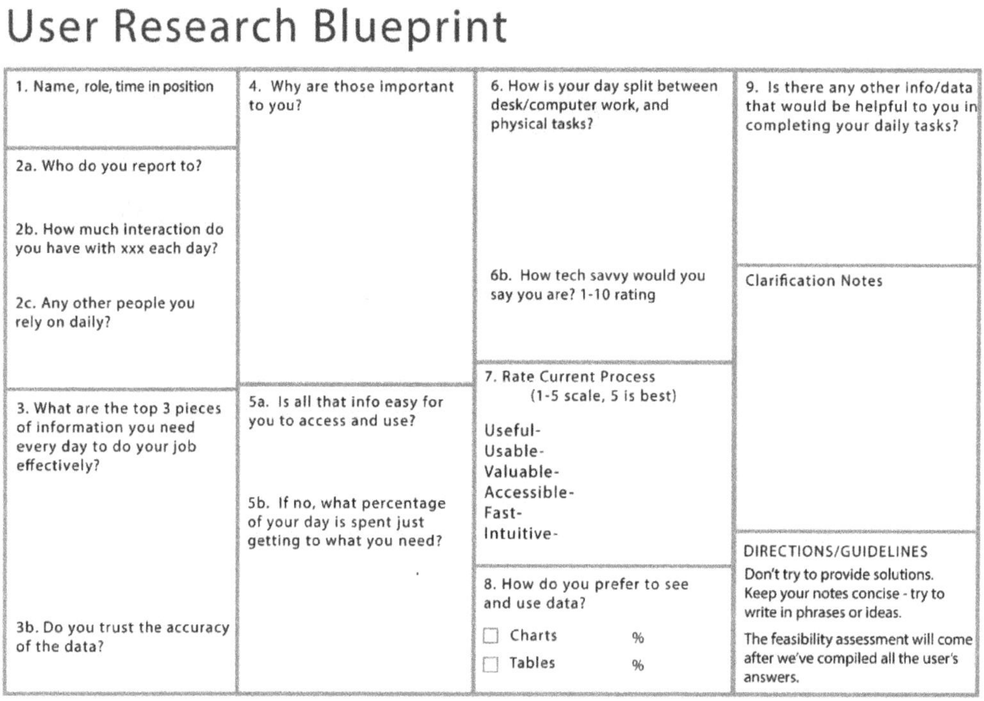

User Research Blueprint

1. Name, role, time in position	4. Why are those important to you?	6. How is your day split between desk/computer work, and physical tasks?	9. Is there any other info/data that would be helpful to you in completing your daily tasks?
2a. Who do you report to?			
2b. How much interaction do you have with xxx each day?		6b. How tech savvy would you say you are? 1-10 rating	Clarification Notes
2c. Any other people you rely on daily?			
3. What are the top 3 pieces of information you need every day to do your job effectively?	5a. Is all that info easy for you to access and use?	7. Rate Current Process (1-5 scale, 5 is best) Useful- Usable- Valuable- Accessible- Fast- Intuitive-	
	5b. If no, what percentage of your day is spent just getting to what you need?	8. How do you prefer to see and use data? ☐ Charts % ☐ Tables %	DIRECTIONS/GUIDELINES Don't try to provide solutions. Keep your notes concise - try to write in phrases or ideas. The feasibility assessment will come after we've compiled all the user's answers.
3b. Do you trust the accuracy of the data?			

Figure 8.1 – User interview blueprint

Imagine that your interviewee is seated. You've introduced yourself and the team and made sure they're comfortable.

Interviewer: "Okay, Joan, we'll get into the interview now. Please know that this is a safe space, and we rely on you being as honest as possible. We really need to understand as best as possible from your perspective how the application works or doesn't work. Sound good?"

Joan: "Yeah, sure, sounds good."

Interviewer: "Great. First, for the record, we need your name, the role you have here, and also how long you've been doing it."

Joan: "My name is Joan Schmidt, I'm an operations analyst on the sales team, and I've been doing the job for just about eight years now."

Interviewer: "Okay, great. This next set of questions will help us establish how what you do fits in with the bigger picture of supporting the company. This is a three-parter, so first, who do you report to?"

Joan: "I report to Frank Miller, the business manager for my region."

Interviewer: "Which region is that?"

Joan: "Oh, sorry, the northwest region. Washington, Oregon, Idaho, and Vancouver."

Interviewer: "Awesome. Part two – how much interaction do you have with Frank each day?"

Joan: "Hmmm. Well, some days, quite a bit, especially when we have a hot problem or a new opportunity. Usually, though, we talk at least once every couple of days, and we have a weekly meeting with him and the other analysts."

Interviewer: "Okay, great. Are there any other people you rely on daily?"

The magic in keeping it conversational and interjecting your personality is that you make the person being interviewed more comfortable, and a comfortable interviewee is an honest interviewee because they feel that they can trust you. No matter who is sitting there with you, or on a Teams call, be yourself and don't be stiff. Have fun with it and everyone else will too.

Next, let's get into the personas and interviews.

Dissecting the stakeholder interview and persona

We'll get into the stakeholder more in the persona part of this section but know that this is a very important person, or persons in some cases, to you and the UX team. It's generally this person that will give you the task, but that's just the beginning. You must interview them to learn as much as you can about the ask so that you can set your UX analytics project up for success.

The stakeholder interview

Interviewing the stakeholder is different than interviewing the users. The stakeholder is responsible for delivering the project to management and, therefore, must be treated differently. They won't always have all the information you need for the best results right away, but through careful interviewing, you'll both understand what you do have and where they need to go to hunt down something or do what they can to support the project success and return on investment for the financial outlay, whether the job is internal or your company is hiring an outside firm.

Here's the blueprint for us to dig into. I've been using versions of this for the past six years with various companies internationally, and it gets you exactly what you need to proceed with your project. Take a look, and then I'm going to explain why each question needs to be asked (you can also download a copy at `https://github.com/PacktPublishing/UX-Analytics-Best-Practices-Tips-and-Techniques-`):

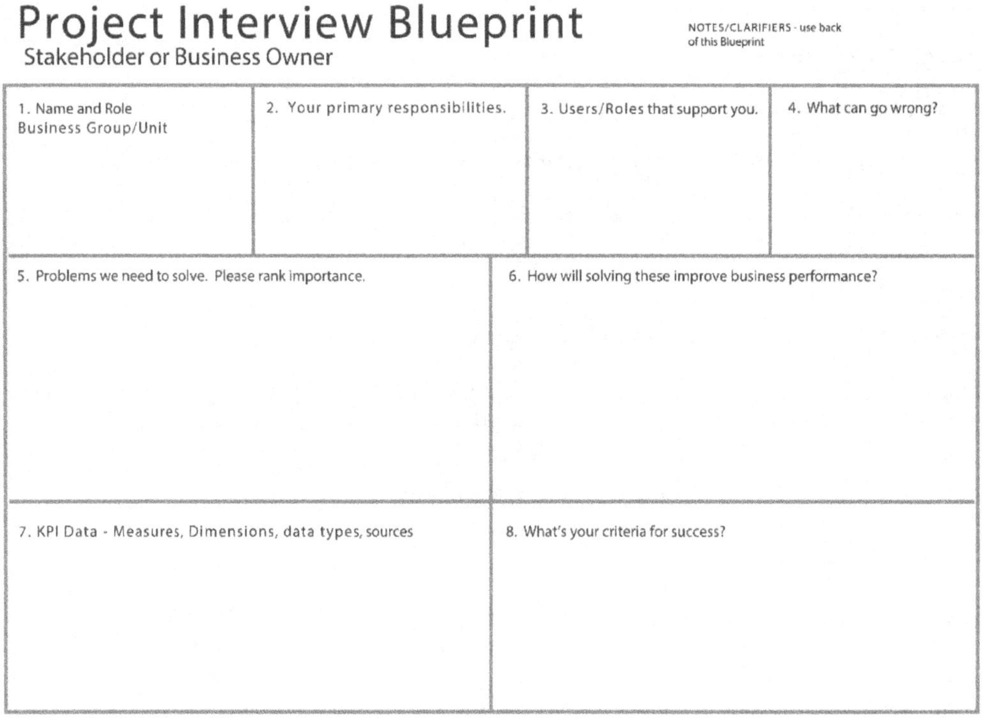

Project Interview Blueprint
Stakeholder or Business Owner

NOTES/CLARIFIERS - use back of this Blueprint

1. Name and Role Business Group/Unit	2. Your primary responsibilities.	3. Users/Roles that support you.	4. What can go wrong?

5. Problems we need to solve. Please rank importance.		6. How will solving these improve business performance?	

7. KPI Data - Measures, Dimensions, data types, sources		8. What's your criteria for success?	

Figure 8.2 – Stakeholder interview blueprint

To set some context first, it's quite possible that you'll interview the stakeholder prior to setting up your team. This may be in person or on the phone, so do your best to get at least one other person to do this with you. That way, one of you will be doing the questioning, and the other can take notes – it's even better if you can get two others. The sequence has a purpose, so always ask the questions in the correct order. So, here we go with the questions and explanations.

Name and role – business group or unit

Name and role is a standard question. We need the role because it helps us figure out where they sit hierarchy-wise within the organization. We get the group or unit so that we know what the funding source is and whether or not other work is being done in the same. Oftentimes, we may know someone else in the group who can help us get the best information possible from this level. Also, realize that the more we ask about where someone is within the organization, the more they realize that we're good at what we do. It may be subliminal to some, but they understand that we're thorough, and that can lead to easier solutions to those hard challenges as we move forward.

Your primary responsibilities

We need to understand what this person is responsible for in their role. Quite often, these people are managers and have a team, but not always. Knowing their day-to-day life at work will start to help us understand more about the functions of the company. This is especially true if you work for a consultancy or a software company and do this work for the customer – not internal to your company.

A tech manager or director in one company could have a very different job description than one in another. Every company is unique, so you need to understand this part of the org structure. And as you continue to do this kind of work, you'll be amassing a database full of useful information on jobs, tasks, roles, and project types. With each new project, you can look back and find comparisons to previous work, and this can help you establish baselines faster.

Users/roles that support you

If the stakeholder is a manager, this is where we start getting the people we'll be interviewing as users, which can be people reporting to them, or supporting them in various capacities If there are a few different roles, then we'll be creating personas from each unless there's been repeat work with the group, in which case you may already have the personas.

If you do have personas, refine them through these interviews. It's possible that with newer technology being used or new managers coming into the business, things are done differently. Personas can and should be reviewed on a regular basis to ensure as much accuracy as possible.

For instance, a stakeholder could hold a position that has managers, report writers, designers, tech people, and so on. If they state in the interview that all these roles support them, grab what personas you already have, and create ones you don't so that you have the clearest picture of the daily life of that stakeholder.

The managers that report to this stakeholder will have people reporting to them, so if you're given the names of managers, ask them to name others that may be in overall support of this project. In larger companies, that could be upward of 20 people. Determine with this stakeholder how deep and wide you should go with individual contributors.

What can go wrong?

This is one of the most revealing questions you can ask in the stakeholder interview. By asking them what can go wrong, you immediately start to identify blockers. We're talking about completing the project, so these could be funding, tech challenges such as getting a new database installed properly and running, losing a key customer before the project is complete, and so on. Knowing these upfront will help you when speaking to users and when presenting the analytics research to the stakeholder.

This question also gives you information on the influence of the stakeholder. If they state a possible blocker, and through the conversation you get the feeling that they may not have the right support to overcome the blocker, you'll need to either figure out a way around it or find a person who can smooth the bump, or assist in eliminating the problem.

Problems we need to solve

These are now the specifics of the project, as opposed to the blockers that might get in the way of proper completion. We're getting to the nitty-gritty here – in the eyes of this stakeholder. Remember that someone else may have given them this task. Or, it could be that they manage a large group and this is their perception. If the project receives the proper funding and go-ahead from higher up the management ladder, then you can feel comfortable that most things will go right.

First, you'll just ask the stakeholder to list what they feel are the top 3 to 5 problems. Then you can go back to rank them. Asking them to name and rank them first can be tough, so make the cognitive load softer by just asking for the problems. Asking for only 3 to 5 also allows them to focus.

It's also possible that this project is the result of real-time, tool-based metrics – those employed and put into place to watch clicks, time on a page, bounces, sources, links clicked, and so on. If this is the case, you should consider two different scenarios:

- You get a complete understanding of those metrics before beginning. This, in my opinion, could introduce bias into your research and therefore keep you and your team from being objective:

  - Because of this, any person on your team could be thinking of a particular metric that they feel is the root of the problem and, therefore, won't be open-minded and unbiased during the interviews. This could render this core team person incapable of being their best. Give this some thought.

- On the other hand, looking at those analytics beforehand could provide clues as to the answers being given by the stakeholder during your interview with them:

  - See my earlier caution and give careful consideration to this option

When considering these two possibilities, I always opt for not looking at the generated UX analytics first. I want them, most definitely, and I may even have another team member study the numbers, but I won't introduce them to this part of the research. You don't look at them until after you've done the stakeholder interview.

This is what I tend to prefer, and even if a stakeholder brings them up, I'll explain the bias part and do my best to keep them separate and out of the minds of my interview team.

> **Caveat**
>
> It's your team that found disturbing metrics and has therefore pushed the need for changes to be made. In these situations, it's your or your team's job to inform the stakeholders of the problems. You already know what they are and will be charged with making the fixes if approved.

Obviously, in this case, you have no choice. You have the proof, inform the stakeholder, get approval, start working on a solution, make the changes, and retest from the new fixes.

How will solving these problems improve business performance?

Now we're getting close to a value proposition. By getting answers to this question, we're understanding the ask at a deeper level. The other questions were asked before this because we set this one up with them. We know what the current problems are, who will be responsible, and who supports all the work involved in the environment and situation that needs a fix.

Answers will vary, of course, depending on the project and problems, but they may be something such as, *"If we solve these problems, we'll be able to open two new facilities by the end of the next fiscal year and increase our revenues by 20%."* Think about this from where you're sitting. What does your company do, and what are the plans and business goals?

KPI data – measures, dimensions, data types, and sources

Key Performance Indicators (**KPIs**) are the numbers that management studies to determine whether the company is doing alright or needs help. If they've come to you with a project, it's because these KPIs aren't lining up correctly. UX analytics – user experience analytics – is the discipline of figuring out why that is. Why aren't the numbers lining up?

Measures are equivalent to quantitative data, and dimensions are equivalent to qualitative data. The sales figures are 15% below where they should be for this quarter. Measure, right? 25% of our survey respondents say that the quality of care we provide isn't where it needs to be for them to refer us to their friends. This is a dimension.

By getting these from the stakeholders, you're understanding company performance standards.

What are your criteria for success?

So, we come to the conclusion of the first round of questions from the primary interviewer with this last one. By asking for the stakeholder's criteria for success, you know what you and your team need to produce. This is the benchmark, so to speak.

You're also now about 20 minutes into the interview, and you've been establishing a rapport with the stakeholder so they feel comfortable with you. Even though this will be more of a hard and fast list of criteria, you have a deeper understanding of this person than you did before.

You've been watching them and listening to how they answer questions. You've been able to guide them through the process with the precise order of questions, and you've done so in a way that gets you and your team the best answers possible.

The criteria for success will allow you to wrap up the interview in a way that lets you and the team understand the nuances and intricacies of the project. This answer will give you a better picture of the people who you'll be interviewing because now you understand what must be accomplished – or proven inaccurate and in need of a pivot or retooling.

The stakeholder persona

Now that you understand the questions and why we ask them, let's paint a picture of this persona – the stakeholder.

Here's what a persona visual looks like. It's from a series of teaching seminars I did in 2015 for a large software company and perfectly illustrates a stakeholder:

Figure 8.3 – A stakeholder persona example

This is a fictitious person based on the characteristics of several people we've done research on – all who hold a similar role with similar needs. It helps us to employ empathy for a person whose job it is to manage at a high level in a company. You can see the categories we need to understand:

- **Skill levels**: By knowing this, we can determine how best to design our research and also how that will influence the design of the user interface.

- **Influencer distribution**: This shows us their decision-making power within the company. If we're supporting a person of this caliber, chances are that our work will also influence the decisions needed to change what our analytics have proven as substandard and pain-inducing to our users.

- **Needs of the user**: We're getting specific now as to what this user – because they are a user, right? – needs in order to do their job efficiently.

- **Challenges of the user**: We see the pains and challenges, so now we have a better grasp on what's currently not right with our software system or tool.

- **Lifecycle of yearly interest**: People at this level work differently than the people they manage. For a top-level CEO, there may be 3 to 8 levels of people below them. That number certainly could be higher in larger organizations with upward of 20,000 people. Knowing when they have a peak interest in seeing metrics is another key element of this type of persona.

We have a quote as if from their own mouth, and on the right of that would be more copy that explains what they do and what they need in the role.

Because we've put a face to the role, it allows us to understand the needs at a deeper level. We've *personified* the role in a way that we can use empathy and compassion in understanding the needs and challenges.

The illustration given is simply a sample. There are many ways to design personas, so feel free to use this as a template and adjust it to your own needs.

Let's now move on to the user personas and the interviews we'll do with them.

Grasping the user interviews and personas

As we saw in the stakeholder blueprint earlier, we get the names of people who support the stakeholder and will be interviewed. Look at the following figure, and then I'll explain the best practice involved:

Figure 8.4 – The sweet spot of understanding

The best possible outcome in 99% of situations is realized when we hit that sweet spot in the middle. It's there that the organization and the individual understand each other's needs and agree on a path forward.

This is why we interview the stakeholder first and then the users or support personnel. We don't act on anything said by the stakeholder – we just listen and do our absolute best to understand and write out the problems and the ask. Now we have the *truth* from the business.

The user interviews

Now we're getting to the level of individual contributors and people who support the stakeholders. It's also possible that this group is much wider and supports other people and functions within the organization as well, as mentioned previously in the section on interviewing the stakeholder.

As before, let's first take a look at a user interview template that we saw earlier (*Figure 8.1*):

User Research Blueprint

1. Name, role, time in position	4. Why are those important to you?	6. How is your day split between desk/computer work, and physical tasks?	9. Is there any other info/data that would be helpful to you in completing your daily tasks?
2a. Who do you report to?			
2b. How much interaction do you have with xxx each day?		6b. How tech savvy would you say you are? 1-10 rating	Clarification Notes
2c. Any other people you rely on daily?			
3. What are the top 3 pieces of information you need every day to do your job effectively?	5a. Is all that info easy for you to access and use?	7. Rate Current Process (1-5 scale, 5 is best) Useful- Usable- Valuable- Accessible- Fast- Intuitive-	
	5b. If no, what percentage of your day is spent just getting to what you need?		DIRECTIONS/GUIDELINES Don't try to provide solutions. Keep your notes concise - try to write in phrases or ideas.
3b. Do you trust the accuracy of the data?		8. How do you prefer to see and use data? ☐ Charts % ☐ Tables %	The feasibility assessment will come after we've compiled all the user's answers.

Figure 8.5 – User interview blueprint

Just like with the stakeholder interviews, the primary interviewer should follow this blueprint in sequence. As opposed to the stakeholder interviews, where in most cases – with exceptions – you're only going to do one interview, with users, you'll do as many as possible. Let's dive into the sections. The name, role, and time in position are openers, with the most important being time in position so that we know whether we have an experienced person in the seat or a relatively green beginner.

You're slowing and methodically uncovering usage information of the application in question, or if you're tasked with supporting the design of something new, this information will point you in the right direction. By getting as clear as possible on *current state interaction*, you and your team are building the case for a new design to alleviate or eliminate problems. Let's move on to the questions that you would want to ask.

Who do you report to? How much interaction? Any other people?

This question set will give you information on their day-to-day tasks and activities as it relates to coworkers and managers. It's possible that we'll get a dotted-line type of relationship with other managers and people at the stakeholder level. Knowing these office dynamics can lead you to other people to interview and will help tremendously as we continue to get to the root of the problem and craft a solution.

What are the top three pieces of information you need every day to do your job effectively?

By knowing what these are, we can use them in analysis against the problems and challenges of the project. If several people state the same information is needed, then we start to get a focused picture of what the needs of someone in this role are.

Do you trust the accuracy of the data?

In my experience, this question gets a *"No!"* 80% of the time. Because of company silos and the proliferation of spreadsheets kept in local hard drives, nobody can be certain that what they're getting is good. My teams and I have interviewed people who have told us that there are days when they spend 40-50% of their time looking for accurate data and fact-checking against what they have. That's a huge amount of time and money wasted for a company.

Why are those important to you?

They've told us what data they need and told us whether they trust it or not, so now we ask why it's important to them. Some people will answer this automatically when they're telling you the data that's important, *"I need the sales records every morning because I have to combine the numbers from all 15 stores and file my report by 10 A.M. every Monday and Thursday."*

Is all that info easy for you to access and use?

We've been leading to this point with the preceding questions, and now we're getting into the user's interaction with the current application. With few exceptions, applications provide people with data, information, and ways to act upon what they can get to. We're starting to get to possible pain points now, and if the previous questions haven't given you a good idea of the answer to these, there's a specific question that will get you there.

If "no", what percentage of your day is spent just getting to what you need?

This is worded specifically to align with the pain of not having easy access to data and information. *"...just getting to..."* carries emotional weight and shows empathy for the user who you know has this challenge. Like some of the preceding questions, this may have been answered partially already. If they started talking about this, just reiterate or rephrase, such as, *"That's got to be so frustrating."* You're aligning and empathizing. Then, *"What percentage..."*.

How is your day split between desk/computer work and physical tasks?

This will depend highly on the type of company you're working for or doing a contract for if you're with a service provider or agency. In many types of jobs in manufacturing, shipping, health care, transportation, or retail, for instance, there will be desk time and up and moving around time. Knowing this gives you a better feel for the work and the responsibilities of the people you interview.

How tech savvy would you say you are?

This can be rated between 1 and 10. This is another question that continues to paint a picture of this person in their role. Younger workers will most likely be up around 8 to 10, and some folks who've been around for a while might not be. If a warehouse worker with few computer skills is told that they have to start using a new computer software program, the likelihood of them picking it up fast is slim, and the training the company will have to put this worker through could be extensive.

Rate the current process

This is the most widely accepted criteria test in the UX world, and you saw an illustration of it in *Chapter 1, Figure 1.1 – UX honeycomb image*. When we've gotten these ratings from a number of people, we derive the averages and can show them in a bar chart, which makes it visual and easily understood.

Figure 8.6 shows an example from a project I led a few years ago. The blue bars are the overall ratings from the tech department, and the reds are all the ratings given by the users interviewed. This disparity happened because no UX analytics research was conducted before the tech group designed the application. After phase one of this work was completed, and the team went back to redesign, all the scores went up to over 45%.

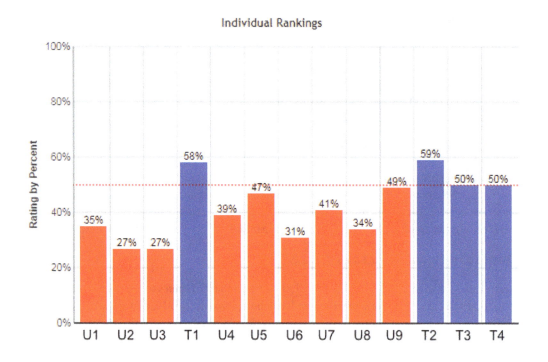

Figure 8.6 – Example of UX usability rankings

The *U's* are users interviewed, and the *T's* are the tech team. When we've done the proper research and visualized the findings like this, it's very easy to put it into a deck to show the stakeholder the results.

How do you prefer to see and use data?

We're trying to find out whether we're interviewing a visual person or someone who just wants to see spreadsheets. I've had people be very adamant, saying, *"Just give me a spreadsheet!"*, *who then* turn around their thinking when I show them the beauty of visual data.

Is there any other info/data that would be helpful to you in completing your daily tasks?

This gets your user thinking a bit more deeply, and in more than a few interviews my teams and I have conducted, the answer has been yes, and there have been several types of information and data to uncover. An answer I've heard many times is similar to this, *"Yes! There's a report that goes out every morning to the senior managers that would really help me advise our VP, but I can't get onto the distribution list."*

So, that was a breakdown of the questions we ask. With just a little practice, you'll be a pro at interviewing in no time. It's always a good idea for you and your coworkers or your core team to change roles, having one be the interviewee and one the interviewer so that you can all get comfortable doing the interviews and gathering the best intel you can for your analytics project.

The user personas

Let's start right off with a user persona for a full-time retail employee. You'll notice that this is designed differently than the stakeholder version but still gives you pertinent information about a person in this type of role.

Estella
Fulltime Associate

I've been working here since I graduated high school. I love my work family and the ways that I can contribute to all the great customers who come to me for advice.

Estella is a working mom with 2 kids - one just starting high school and the other in kindergarten. Her husband owns a small construction company in Beavercreek and together they remodeled their home.

Spanish is her first language but she speaks English fluently as well so she's a great asset in a store that caters to a diverse community of shoppers.

Although she sometimes struggles with learning new tech, she's a very realiable worker who does her best to help her fellow employees and is a star with customers.

She can be counted on to get her work completed every day, to the standards expected, and her managers always see her as a key player on the sales floor.

PERSONAL INFORMATION

Age: 32
Gender: Female
Education: High School
Hometown: Dayton, Ohio
Status: Married, 2 kids
Languages: Spanish, English

NEEDS OF THE USER

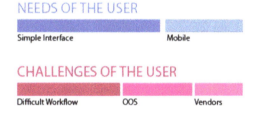

Simple Interface   Mobile

SKILL LEVELS

Technology...............
Communication.........
Leadership................
Sales............................

CHALLENGES OF THE USER

Difficult Workflow   OOS   Vendors

Figure 8.7 – User persona example

These are usually easier to create because you interview many users and can get the best composite of this worker. After creating this persona, it's reusable any time you work on this type of analytics project. If you work for a company that serves various industries, you'll soon have a folder full of personas that you can bring out and use any time it's needed.

Sometimes user personas will have to be fabricated. You might be in a position where there's no budget or time for interviews, and you need to come up with something. Based on the stakeholder interview and what you uncover there, you can create the one or two that are needed so that you can put a face to a user and can use that for the empathy and compassion needed to lead the design team in the right direction for the application build.

Now we're going to move on to the actual interviews. They're really quite simple but do have some strict rules and codes of conduct. If followed, you'll build rapport with your interviewee faster and get the most honest answers they can give.

The importance of interviewing one person at a time

I'll do my best to relay the extremely important nature of this facet of your UX analytics work. There are two main reasons for this:

- There will be times when an interviewee wants to tell you something but is afraid or reluctant to do so in front of other people who they work with

- There's always a stronger personality who will answer the majority of the questions when interviewing multiple people at once, and you won't get a fair assessment of the problem

The interview blueprints are designed this way on purpose as well. Each blueprint represents one user and their answers, pains, and challenges with current state workflows. If you've got the answers of two or three people on one sheet, you can't keep things separate and independent.

Take this advice from me – with 20+ years' experience in the UX field, working with the largest corporations in the world and hundreds of people in many different roles, you don't want to break this rule.

It's especially important to not have a supervisor in the room when interviewing. You'll have people say it's okay and that everything will be fine. In 99.9% of cases, that's a false assumption. If you state the reasons with conviction and explain why we – and the best companies in the world teach it and do it – you can, more often than not, make sense to them. If this fails, take it up the ladder to someone with a higher position and more authority in the company, such as the UX director or a UX champion who believes in the mission of UX.

Sometimes things don't work out as planned

There will be times when an interview goes off the rails or veers in an unintended direction, especially if you absolutely can't interview one person alone without a manager or in cases where you're given two or more people at once.

Fear not. In the case of, let's say, a phone interview that was set up for you, and you weren't able to specify the role needed, you want to start with what we call "knock-out" questions. These are qualifiers that, when answered in a specific way, the interview ends. These are also very good for doing mass surveys if you're paying for the service and don't want to waste an entire run-through with someone who you don't need to be interviewing.

Example question with a yes/no answer only: Are you involved in handling products at a loading dock on a daily basis? If the answer is no, game over and the person exits. If they answer yes, they move on to the rest of the interview or survey.

If you do an interview that gets sidetracked, don't despair. These things happen, and my teams and I have decided that all information should be kept, recorded, and stored away because you never know when even the slightest bit of information from an interview could be useful.

The reason why I use the blueprints with specific questions, rather than just freestyle, is to help mitigate these types of problems. If you stick to the plan, it's much more likely that you'll get valuable insights from your interviews. Be mindful of the answers you're getting and steer interviewees back on track as soon as possible when they start down paths that don't serve your purpose.

Summary

This was an intense chapter, and though there weren't very many subjects covered, it was in-depth and detailed. You learned about the two main interview types, of a stakeholder and a user, with users generally supporting the stakeholder. Each is quite different, and from each, you get a unique perspective on the business and the people who support it. The sweet spot illustration in *Figure 8.3* helped you understand this.

You saw what a visualization from a research session looks like and learned the importance of showing these to help decision-makers make the best choices. And for each interview type, we covered an explanation of why we ask the questions we do and in the order they're designed. Each builds on the next and creates the best analysis of that user type or stakeholder.

In *Chapter 9*, we'll switch to research that's not conducted in person – the survey. These present their own unique challenges, and getting solid on these best practices is key for your project if surveys are part of your arsenal.

9

Surveys – A High-Level, Broad-Stroke Tool for Quick Direction

Now we will move from the interviews and personas involved, as covered in *Chapter 8*, on to a broader research-gathering technique – surveys. Sometimes, it's necessary to move out into the market and get a broad sentiment of the problems you're trying to solve. Surveys are a way to do this quickly and more efficiently than is often possible through in-person interviews.

Budget will be a concern here, as the best tools to use aren't free and can be quite costly when you need to get responses from hundreds, or even thousands, of people. They can be quick and help set the focus for further research if the questions being asked are well thought out and vetted with your entire team. You should only use surveys after the stakeholder interviews are complete and the positive direction is agreed upon.

Sometimes – and this really is a case-by-case basis – it can be helpful to do a small sample size survey before doing your user interviews. If you have the budget to do a 50 to 100-person test, with questions based on the previous research from your stakeholder interview, you can get market sentiment if, and only if, you can be assured that the interview cohort can be focused on the particular user personas you will be designing for.

In this chapter, we will focus on the following main topics with respect to surveys:

- The audience
- The questions
- The data

Defining your audience

Surveys are extremely tricky if you can't identify an audience. With any good survey tool, you need to start with what's referred to as a *knockout question*. This will ensure that the survey is only taken by people who fit into your intended audience.

Look at your audience in terms of personas. Those personas are the people responsible for using your product to do the jobs they're tasked with. This part can tie into what's coming in the next section on writing the questions and sampling surveys. If you already have a good idea of the user, as in they are employees of your company and you already know who they are, then you can work from those personas.

Regardless of who your intended or actual users are, make sure you get clear on this in your stakeholder interview. Trying to do surveys without this understanding is the recipe for failure.

Here are a few examples of audience definitions to get you thinking:

- Shipping managers for warehouse distribution centers. They have to ensure that the right product gets put on the right trucks, going to the right customer.

- Hospital ward managers who have to control inventory, manage staff, and ensure the overall quality of service for their part of the institution.

- Customer service reps who work for college loan servicing companies. They must keep track of their students and help counsel them on repayment options while also understanding everything about the loan and lender.

Knockout questions

As stated in the preceding section, knockout questions eliminate people you don't need right away. If you need mechanics, you don't want file clerks. If you need file clerks, you don't want doctors. These are overly dramatic examples to make a point: you want to only spend your credits on people who most closely fit your persona.

With survey tools that have a team behind them, such as **Maze** or **Qualtrics**, they have cohorts of individuals that they dip into for the research – these will act as your demographics. You can specify upfront the direction you are heading in, and they will set you up with the closest group of people they can. Once that's established, your knockout question becomes a further filter for getting to just the people who can answer the remainder of the questions and get you moving toward the analytics metrics you need.

Always keep in mind that until you start your customer research, and sometimes even after you start and refine your scope, your audience can change. For example, if these surveys lead you and the team to believe or understand something other than what they did when you started, you may have to pivot and rewrite this audience persona. This can also be a factor of ambiguous findings, where not much clear direction is coming through. Anything that's fuzzy and hard to turn into a solid hypothesis should cause you to reevaluate and make the changes needed.

Writing and rewriting the questions

While the knockout question is intended to whittle out people who don't have the job or tasks you need information on, the other questions are written to get you closer to the needs of the market, and that understanding can help you with your user interviews, whether internal – most ideal – or external, where you're interviewing people not within the walls of the company you're gathering analytics for.

So, let's explore questions, from knockouts to detailed, that will illustrate how to start super specific, then go wide for a bit to start directional awareness, then refine back to specific for the finer details that will inform your teams for the solution discovery, which comes after your work in research is complete.

Writing opening-level questions

You'll first want to write questions that let you know whether you're headed in the right direction or are off course in some way. For example, if the people who pass through your knockout question give you answers that don't support what your stakeholder-approved direction is, then you have some thinking to do and a possible direction change. You can, at this point, do a handful of user interviews with the same questions to see what you get internally, but if your personas match, you've got a good indication that your project isn't ready for the next steps.

> **Important note**
>
> If you're doing work to fix problems with the adoption of a custom application, one that isn't used in the market, then you have to stay internal with your users. Surveys will only help if you can ask questions about workflows with market-available applications.

Let's work from a scenario, and we'll make it simple so that it's easy to understand and use as a foundational piece of your work.

Here is the scenario. Your company just purchased new business intelligence software and the phase-one user group is sales analysts. They've been using the software for 60 days with little success, and our stakeholders are telling us that only about 30% of the intended users are actually making use of the tool. In order for the ROI to prove out, that number needs to get up past 75%.

Okay? In this scenario, it's a **commercial off-the-shelf** (**COTS**) product that's used by an estimated 200 enterprises globally. So why are your users having trouble with, and failing to use, the new application?

Since this is a commercial product, not one designed and built in-house, a quick survey is what you can suggest to get market sentiment first, and then go to your internal users. A survey can get you results in a couple of days, whereas interviews might take a few weeks to schedule and find the results. The interviews can also be more costly than surveys, and you don't have that much time to find direction. If you have budget constraints, that's another nod toward doing surveys.

And for now, while just getting used to how to write survey questions, we're only going to look at the questions, not how we design them in the survey software. We'll get to that later in this chapter with more explanations.

The knockout question is, **Do you use XYZ sales analysis software in your job?**

If **Yes**, this user will continue; if **No**, they will be thanked and that's the end for them.

Assuming **Yes**, the next questions will be as follows:

- **How long have you been using XYZ software?**
- **What percentage of your day is spent using XYZ software?**
- **On a scale of 1-5, with 5 being best, how would you rate the ease of use of the software?**
- **Does your work with the software go upstream to managers, downstream to other workers, or both?**

Survey complete; they get a thank you for participating, and you've got a start with UX analytics metrics and data.

If you got 35 out of 50 that finished the survey, you have some decent data to work with. The first question gives you an idea of whether or not someone is skilled at using the software or is a novice. If, for instance, you have a number of skilled users who rate the software at a 2 for ease of use, then you know you've got some difficult software on your hands. The question about the percentage of the day using the software gives you an understanding of how pervasive the application is in day-to-day business for that persona. By asking who the results of work go to, you have an idea of where to look for bottlenecks in the overall workflow of a business unit, while understanding their position in the company.

If we assume, for the sake of argument, that this survey gives you a strong indication that it's not just your users who have difficulties with the software, the report back to the stakeholder needs to support a position that will cause management to make what could be a difficult decision. They will either put the money needed into solving these problems or just eat the investment dollars and find a better solution.

Either way, your UX analytics work should point out as clearly as possible what the sentiment is, how the problems are defined, and, if you also are tasked with possible solution discovery, make recommendations on what to do next.

The flip side of this scenario is that your survey answers indicate that those users outside of your organization really like the software, use it every day, and support upstream work that enables company growth. If this is the case, your problems are internal, and your job now becomes focused on what those internal problems are.

Refining and rewriting the questions

Staying with our scenario, we asked these questions:

- **How long have you been using XYZ software?**

- **What percentage of your day is spent using XYZ software?**

- **On a scale of 1-5, with 5 being best, how would you rate the ease of use of the software?**

- **Does your work with the software go upstream to managers, downstream to other workers, or both?**

And we'll assume that the results of this survey show us that the market is in favor of the software. Here are our averages:

- How long using? Two years (which would indicate proficiency).

- Percentage of day? 60% (high daily usage).

- 1-5 ease of use scale – 4 (an average; some people rated at a 5, some at a 3).

- Which way does work go? 75% downstream, 25% upstream.

Here's the narrative for this analytics outcome:

"Our survey of market acceptance of XYZ software has shown a strong positive. With an average usability and acceptance rating of 4 on a 1-5 scale, and the average percentage of the day spent using the software at 65%, we can safely say that it's a solid performer. Our suggestion, with our current understanding of the data and the intentions for its usage within the corporation, is to interview internal personas to find the challenges and pain points our users are experiencing. From there, we can work with the stakeholder to determine the next steps, which could result in a slight pivot of direction and roadmap."

This should be part of your report along with the visuals from the data and any other supporting data and information you may have. Let's move ahead now to the refining of the questions for a deeper dive survey.

Those questions gave us great information on how a number of other companies and users liked the software, but we do need to go a bit further. Since the market is in favor of the application, we know that something is amiss with our users. This is the point at which we must make the decision to either do another survey externally or bring it in-house to our users.

Staying external for further refinement

If we choose to do another survey externally, what more do we want to learn? Okay, you say, we know that the average tenure in using XYZ software is two years, so what does that really tell us? And what if the work distribution were the other way around, 75% upstream and 25% downstream? That could change things dramatically.

Let's write some refinement questions and go back in with another 50 to see what we get. Notice that we'll start with the original questions, and then dig down a bit. Depending on the survey service you use, this can go a couple of ways – I'll show you one that we prefer in our work currently.

How long have you been using XYZ software?

1. **1-3 months**

2. **3-6 months**

3. **6 months to a year**

4. **1-2 years**

5. **More than 2 years**

The answer to this question can fire off branches of next questions. For instance, if they answer **1-3 months**, the clarifying question will be different than if they answer **1-2 years**. Because we're getting the idea that it may take more time for people to get used to the software, and to use it efficiently, we really want to dig in here. Let's refine the people who've been using it for a shorter time.

Since you indicated that you've only been using the application for 1-3 months, how proficient do you feel?

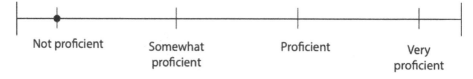

Figure 9.1 – Likert scale on proficiency

Now we're getting somewhere. This answer will give us a huge understanding of the ease of use of the software. If the average rating is **Not proficient** for people who've just been using it for 1-3 months, we know that the learning curve may be about normal. But if we get that same rating from people who've been using it for 3 months to a year, then we know there's a steep learning curve and our users may in fact be the norm for the application.

What percentage of your day is spent using XYZ software?

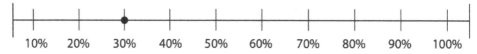

Figure 9.2 – Likert scale on percentage of day

If we get the **Not proficient** answer as in *Figure 9.1*, and then an answer of **30%** in this question, we start to figure out that this is an application that needs to be used regularly during the day to get proficient. So, the clarifying question for anything under 50% could be as follows:

If you were to use the application more frequently throughout your day, do you think your proficiency would increase rapidly?

○ Yes
○ No

Figure 9.3 – Radio buttons

Well then, what happens if the respondent answers **No** to this one? And further out from that, what if the majority of your respondents answer this way? They're telling you that more time won't make a difference – it's a very difficult piece of software to understand and learn. We should probably add a clarifying question if **No** is the response. Maybe like this, with a freeform textbox that they can write in:

What do you find most difficult about using the application to accomplish your tasks? Please be specific.

This question will get us direct answers on why these users feel that the application is unusable. When this is synthesized, we'll have external indicators that can be compared to internal sentiment from interviews we'll conduct with our users. In essence, we're validating our findings in both ways to give the best information to the solutions discovery team.

On a scale of 1-5, with 5 being best, how would you rate the ease of use of the software?

Figure 9.4 – Ease of use scale

This indicates very poor ease of use and supports the direction of the **No** answer. Following along with this scenario story, the average time spent in the application is only 30% per day and the average experience with it is 1-3 months. What can you infer now? Are you starting to feel that, given time, your users might catch up and get over the learning curve hump? Or do you have another thought? Jot down what you think so you can come back to it later.

The **Yes/No** question about the learning curve – where we're assuming a **No** answer for the scenario – gave us what we needed, and the follow-up question helped us understand why the respondents answered that way. We want them to think through their experience and ask the questions preceding to see then how, after doing that thought process, they rate the ease of use. While it would be very unlikely that we'd get anyone giving us an answer over 3, considering the previous answers, it could happen, and that would necessitate another direction change or see it as an outlier if it's only a couple of answers out of the many that you got.

Does your work with the software go upstream to managers, downstream to other workers, or both?

Remember our change on this one – we're turning it around to 75% upstream and 25% downstream. Think this through for a minute or two before reading on. How does this change your perception of the application and the personas? How does it change the way you think about the ask from the stakeholder? Give it a couple of minutes before continuing; it will help you start to consider your options and think like a UX analytics professional needs to – strategically.

Okay, let's continue. With the turnaround to 75% upstream, you can consider this to be more of a management tool than a tactics tool. In other words, it's a tool used by the enterprise to capture data for decision-making at a higher level. It's more of a business planning tool than a day-to-day execution of tasks tool.

Another version of a Likert scale can work here, with a straight sliding scale for the respondent to interact with. This one shows weighting toward management.

Figure 9.5 – Task versus management orientation

Think back now to the other metrics we've gathered and what they showed us:

- **Not proficient**
- **30%** of the day spent with the application
- **2** on a 1-5 rating of ease of use
- **No** on more time to learn

Here's what I can imagine a narrative of this might sound like: "*Yeah, I'm not proficient with this tool at all. Maybe it's because I only use it for about 30% of my day, but I don't know. I've used other applications that do similar things and they were much easier to use than this one. Rating? Mmmmm, I'd give it a 1.5 to 2 honestly. Almost half of my day is spent trying to figure out what to do. There are so many options to choose from in some steps and most seem irrelevant to my work. I just don't get it. I doubt whether spending more time during the day would do me any good. Like I said before, half my time is trying to figure out what's going on.*"

There comes a point where people get so frustrated with trying to use a new tool, application, or system that they just give up. Even though their old way of doing things may be antiquated and mostly manual, they know how to do it and it's comfortable for them.

Taking all of that into consideration from earlier, you should be realizing that there's a definite problem somewhere and that your next step will be to start interviewing people internally if you used these surveys to help set up those steps.

Gathering and synthesizing the resulting data

You've completed the rounds of research analytics you and your team feel is needed, and your stakeholder has agreed that you're in a good place. If there's a project manager involved, they have been following along and also agree that you can move forward. The gathering and synthesizing isn't something you save until the end – you'll be doing this throughout, and then putting it all together for your readout.

In this phase, you need to produce two types of information: reports and visualizations. The reports should be written in narrative style and should contain all explanations of your work. Include the conversations and decisions you and your team came to from working with the stakeholder. Include the hows and whys of all the research types you've done from the surveys to the user interviews.

Document everything down to the last detail so that anyone can read and understand. Document everything so that, if you have to support any of your claims, you have proof of what was done.

Here are some of the particulars:

- The aggregate answers from all your surveys, meaning that you need to document the averages from all the survey answers and have those numbers in the report. Where you have qualitative answers, give the average answers there as well. State the question as it was asked, and then frame the answer such as, "70% of respondents indicated that they felt the workflow contains too many steps."

- A spreadsheet with all your numbers, whether quant or qual. Remember that you can indicate qual with percentages. Even if the ratings you're using are word-based – **Not Good**, **Good**, **Great**, and **Fantastic**, for example – put these in the spreadsheet too. With a column header that indicates to you and your team what question the answers belong to, it's easy to refer back at any point in time. These spreadsheets are what you'll use to feed the data visualization.

- A data visualization or business intelligence tool such as Power BI, Tableau, or Qlik. There are others, so if your company uses something else, go with that. While there are many ways to use data visualization, to get a start, concentrate on simple graphs that you'll reference in the documentation.

This last bullet is very important and so we're going to cover visualizations in detail in the next chapter. So, let's wrap up this chapter and rehash what we learned in it.

If statistical significance is required

In some cases, your stakeholder may require you and your team to prove statistical significance before a judgment is made on redesign costs. Statistical significance helps determine whether the results observed in a UX study are likely due to chance or whether they represent a meaningful and reliable difference.

This will most often occur when determining whether or not a new product offering has merit and a chance for sales in a particular market. The bottom line is always whether we will get a return on our investment in design and development spend through sales and service of the product. Companies don't like to spend money on ventures that won't produce a profit. In other words, if nobody wants it, why should we build it?

Summary

This chapter on surveys gave you both a broad perspective and a finer look at the intricacies of using this type of UX analytics to prove or disprove the direction set by the stakeholder. You learned that it can be used to first establish a direction for the user interviews or to further prove the market sentiment to strengthen the position you've already found in your user interviews.

You saw a bit about different ways to use survey tools to allow people to answer the questions in a logical fashion. When looking for answers along a sliding scale, you can use **Likert scales**, which make it very easy for respondents to create a visual picture in their head, and give you the best answer quickly.

Most important, though, to you is your ability to quickly decipher answers and attitudes with very simple questions to start, and then refine for iterations so that you get to the real core of the problems that your team is facing. Sometimes you'll find supporting evidence, and sometimes you'll find evidence that refutes what's happening internally, or with the client for whom you and your team are doing the work.

Next up, we'll continue with something touched on previously, but in a way that explains them as operational measurements – quantitative and qualitative metrics.

10

UX Metrics – The Story of Operational Measurement

Operational measurement is how companies test work and brand recognition to gauge success. **Operations Expenditure (OpEx)** is the money they spend to grow the business via talent, conferences, marketing, advertising, and so on. It's everything they spend that isn't on machines, equipment, rent, and so on. If we as UXers can grasp the significance of this, we can be huge supporters of companies' goals for growth, reduction of costs, and curtailing of losses due to inefficiencies and poor design.

For companies that are striving to serve customers with the highest quality while continuing to grow brand awareness and revenue, using UX metrics in an analytics practice is foundational. Moving on then from our discussion of surveys, in this chapter, we'll show how we can use the survey results, along with our other research and analysis to support our company in achieving those two things.

We learned what qualitative and quantitative metrics are in *Chapter 3*, and in *Chapter 8*, we learned about different personas and how to interview them. Now, we're going to open all that research and expose everything we've learned so that you can do the following:

- Understand how these two metric types are essential to companies

- Know when to employ each type and see their relevance in the interviews

- Know which visualization methods are best suited to each

- Understand some of the basic math involved

The goal for this chapter is to start bringing pieces together and further solidifying your knowledge on the best practices for UX analytics research and discovery. Nothing, and I repeat, nothing is more important to you, your career as a UXer, or your ability to support your company's goals, than your expertise in pulling together and synthesizing many different types of data and information into actionable steps for your company to follow. So, let's get moving.

Quick review of quantitative and qualitative metrics

This is not just a review, but a true UX professional's way of looking at the world. You and everyone around you are bombarded every single day with thousands of sensory inputs. In research I did for another book I've written, it's estimated that we now receive information so fast that we're forced to make as many as 30,000 decisions a day.

And here's the kicker: all of that – all that input and all those decisions – are either quantitative or qualitative in nature. Yeah, that's right! One or the other, both at the same time, one dominant over the other... Think about that for a minute and let it sink in. Your work in UX analytics can and will determine the paths that allow people the luxury of making decisions, quite often critical decisions, more easily and with greater confidence.

There's no denying that we live in a digital world now and in the years to come, we humans will get more and more entrenched. We're seeing the blossoming of AI in ways we couldn't have imagined even 5 years ago. Who would have thought that Chat GPT and Jasper could write books, or that visual AI could generate NFTs that sell for hundreds of thousands of dollars? Who could have imagined the blockchain?

Somebody did, of course, but those people are rare individuals and make up a very small percentage of the human population. All of the rest of us out here now make use of what they imagined on a daily basis. Our phones, appliances, and cars are all smart now. What you have to decide, as a UX professional, is where you intend to be in all of this. Notice I didn't say "want" or "hope" – I said "intend."

Quantitative metrics – counting things and inferring quality

Because we're going to cover visualization techniques in this chapter, we're going to start thinking visually now. The illustration in *Figure 10.1* is how we'll start. Take a look, and verbalize – or grab a notepad and write – what you see. Be as specific as possible and please don't continue until you've done that. I want you to understand this at a gut level, not just cursory or scratch-the-surface. I want you to dig and work for it.

Figure 10.1 – Visualization example 1

Five squares. Five gray squares. They represent first-level visualization, but there's so much to glean from this super simple illustration. I sincerely hope you wrote down what you see but whether you did or not, here's a way to understand this illustration.

I see five gray objects that appear to be squares. They also appear to be evenly spaced horizontally, and aligned through the center. The color is a medium gray. Since I can count them, I'm using quantitative analysis. If this gray color is specified as the standard, I can attest that they all appear to be the same color, and therefore I can infer that I'm using qualitative analysis as well. If the spacing is a set standard, that's then qualitative, as is the standard size. If five is the correct number, then my counting in qualitative analysis is also correct.

If I'm analyzing this for accuracy, I can say that both quantitative and qualitative analysis is verified for accuracy.

Fairly simple, right? But I didn't just say I see five gray objects; I explained everything I inferred from the illustration.

Now, let's change things up a bit. The following illustration must conjure new thoughts and inferences from you, the UX analytics team member whose job it is to check accuracies and report inconsistencies or raise questions that cause others to check against standards.

Figure 10.2 – Visualization example 2

What do you notice now? There are still five squares, and they all appear to be the same color. Write it down to get the maximum understanding from these lessons – it's crucial for your success.

We're noticing now that something changed, so how do we articulate that? What's the new story we need to tell? If the standard is what we saw in *Figure 10.1*, then something's off with this one. Let's write it out.

I see five squares that all appear to be the standard gray color. However, counting from left to right, square 4 differs in size from the others. It's larger and appears to align at the top, rather than along the horizontal middle axis. Since it's larger, the spacing is now uneven between 4 and 3, and 4 and 5.

Question: Is it a different illustration that means something specific, or is it a mistake that needs to be corrected?

You've seen something that doesn't seem to fit, so you're recording and questioning. And here's where it gets more involved: users of your application, whether they be external customers or workers within your organization, are asking the same question. if there's no up-front context, they are first internalizing it at a subconscious level, and then interpreting it.

I teach this same lesson in all my consulting classes. I teach it to executives and engineers. I teach it to developers and managers. Everyone needs to understand these concepts and you most of all, UX professional, need to understand and master not only the 'seeing,' but the 'saying.'

Here's another one, and if I were sitting right next to you, I'd want you to say to me what you see, what you can infer, and what new questions you have about it:

Figure 10.3 – Visualization example 3

And another one, switching it up a bit more:

Figure 10.4 – Visualization example 4

And the last one for this exercise:

Figure 10.5 – Visualization example 5

These last two head in a different direction than *Figure 10.1* or *Figure 10.2*. Now, the pattern is uncertain, and because there's a blue square and an orange circle, there's doubt as to what you see and what you can infer.

And yet, you need to be able to explain it and ask questions that will require an answer from someone else.

There are five objects in this array, arranged along a horizontal axis – three gray squares, one blue square, and one orange circle. The exact arrangement is: gray square, blue square, gray square, orange circle, gray square. It could be said that the gray squares are the odd numbers: 1, 3, 5.

The objects appear to be evenly spaced and the orange circle appears to have a diameter that matches the length of the sides of the squares.

Question: What does the blue square signify?

Question: What does the orange circle signify?

Question: Why is the blue square in position 2, and the orange circle in position 4?

I've put all this in the subheading for quantitative analysis – things that can be counted. Why did I do that? There are five objects in each example. Is there anything you can infer from that? Notice that each example can still act as a quantifiable reckoning. Five squares, four gray squares and one blue, three gray squares, one orange square and one blue square, three gray squares, one blue square, and one orange circle.

Ready? Three births, one car accident, one open-heart surgery before noon today. Four students with an average of over 80% and one with an average of over 90%. Four basketball players under seven feet tall, and one over seven feet tall.

Again. Four apples in the basket and one orange. Three steaks, one pork roast, and one whole organic chicken. Three pairs of jeans, one belt, and one white button-down shirt. Four size XL t-shirts and one size M t-shirt.

Again. Three people clicked the green button, one clicked the red button, and one clicked the gray button. Four people made it to step 6 with no difficulties and one got stuck on step 6. See where this is going?

> **Important note**
>
> Everything in UX analytics is controlled by context. If the directive is to support the sales team by constantly analyzing time spent by analysts in the sales application, producing reports, and giving strategic suggestions – that's qualitative. If the directive is to provide all the sales figures from each analyst in every region – that's quantitative.

Visualizing research results for maximum influence

Now, we're going to get into more complex analytics visualizations and walk through understanding and using these mechanisms. While it is true that there are different learning styles, we are all visual creatures. We notice when 1 widget out of 100 is a different color or size. We notice when we walk into a room how people are grouped, and who doesn't quite fit the mood.

Therefore, we need to give visual representations of our findings so that decision-makers can come to the best conclusions faster and with greater certainty. If we're counting how many people use an application each day, separated by the times of day when they're most active, we can use a bar chart. This would then be an example of quantitative metrics:

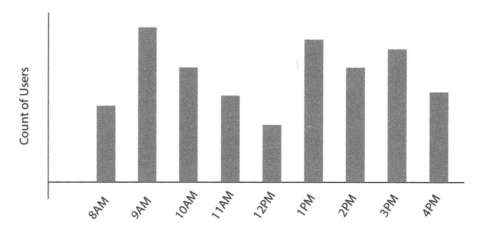

Figure 10.6 – Bar chart distribution example

When you look at the preceding bar chart, what questions jump to your mind? How about *Why are 8 A.M. and 12 P.M. the slowest?* or *Hmm, I wonder why 9 A.M., 1 P.M, and 3 P.M. are high, but there's a dip at 2 P.M..* These will most likely be questions that stakeholders will want to get answers to, so as the UX analytics pro, you must think this way also. And the thinking should lead to the doing – finding out why this pattern occurs. Which, of course, will lead you to qualitative analysis – right?

Every quantitative metric, whether it shows favorable or poor results, should lead to a few obvious questions:

- Why is this happening?
- What causes this type of pattern?
- How can we stop it?
- How can we replicate it?

And what types of questions are those? Qualitative questions! Do you see what's happening here? Quants lead to quals! Yes, they do – always. If you sell 15 NFTs for top dollar, why were you able to do that when 95% of others have no clue? If you have 47 pairs of sneakers in your closet, why? If it takes you an average of 23 minutes to drive to work, why do your coworkers who live in the same area take an average of 32?

You're getting the big picture now, I can tell. Let's keep moving and talk more about influence, the superpower that UX analytics gives you when done correctly.

Donut charts for capturing qualitative metrics

There are too many visualization styles to cover in this book, but let's show a quick example of a style that's good for qualitative metrics – the donut chart. In this example, we see percentages of sentiment across all respondents. If it's the application we're tasked with fixing, we know from this that the majority of people think it's an okay application. A very small percentage say they love it, and about four times as many say they don't like it at all. Those two would be your outliers with a total of only 22% of respondents.

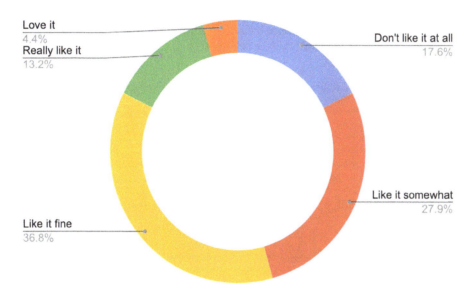

Figure 10.7 – Donut chart example of ratios

I think you'll agree that a chart such as the one in *Figure 10.7* is much more impactful than a table showing the figures, like the following:

Sentiment	Number of Respondents
Don't like it at all	12
Like it somewhat	19
Like it fine	25
Really like it	9
Love it	3

Figure 10.1 – Tabular data example

You can see the results, sure, but there's nothing dramatic or eye-catching about a table full of numbers. Visuals allow us and our audience to internalize the analytics more deeply and with stronger conviction. Always incorporate data visualization into your readouts and if you want to put them into a deck that will accompany your document, that's always a good idea.

So, what about this chart and the numbers it portrays? What is it telling you about your application and the people who use it? Let's break it down and figure out what to do next for this UX analytics use case.

> **Important note**
> Operational excellence is our goal. These measurements are what will lead us in the right direction with proper support from everyone involved.

Take a quick peek back at the visualization in *Figure 10.7* to keep it fresh in your mind. Then consider the following results:

- 4.4% of users claim that they love the application.
  - Why such a low percentage of our total user population? What is it that they see in the app, or the way they work within it, that would cause them to rate it so highly?
    - What questions should we be asking them in order to find out?
    - What specific roles do they play in the company?
    - What is the average length of time that they've been using the application? (You remember this from previous lessons, right?)
- 17.6% of users say that they don't like it at all.
  - While this isn't terrible, why are they saying this?
    - What questions should we be asking them in order to find out?
    - What specific roles do they play in the company?
    - What is the average length of time that they've been using the application?
    - How much time are they wasting during the day trying to figure out what to do?
    - Given this response, how often are they actually using the application?
      i. Have they given up completely?
      ii. What can we do to rectify this situation?

There's no need to get into the other percentages – I think you get the direction you need to go and can infer the others on your own. Relate all of this to your own current situation, and to interview questions you may be asked if you're looking for work or asking for a pay raise.

Utilizing 'why' and 'how' in managerial decision-making

We're doing the job of UX analytics, and this job is a very important one for any company that intends to grow in the market. I've said this in several places in this book so far, but it bears repeating. If you can learn to think of your job as expansive, all-encompassing, and critical to success, you'll then also start thinking like a person who is immensely important to an organization.

I want you to embody the spirit of a person whom a company can't live without. And I'm not talking about arrogant behavior; I'm simply stating that you need this mindset to excel. UX is still a bit of a unicorn in business, albeit a tiny little halfling in most companies. So don't think like that – think like a huge, powerful unicorn that can lead and triumph.

Why do I start this way? Simple. You have to deal with management in order to make things move and shake. If you approach these projects with anything less than total confidence in the research you've done and the recommendations you're making, you won't be taken seriously. I've been there, done that, and have the battle scars to prove it. And the t-shirt – I have the t-shirt.

Management needs direction – that's what they're paying you and your team to provide. Somebody stood up one day and said, *We need to be doing UX analytics!* And after a ton of questions and some back-and-forth and explanations and such, somebody else finally said, *Okay!* So, there you are, doing it and pushing it uphill with the hope that a manager with street cred is paying attention.

Everything, everything, everything that management does is based on finances. If you work for a publicly traded company, shareholder sentiment and Wall Street are the determining factors to roughly 85% of all decisions made. If you can prove that the blue button gets clicked 65% more often than the green button in A/B testing and that that click translates into 45% more revenue per month, then management is going to direct the blue button and only the blue button to be used from this day forward.

Of course, it's not all that simple, but it makes the point clear. Management relies on what you do to make decisions. If you've done the work to ensure that bias doesn't weigh into the equation – and this can be a problem – then you're providing the much-needed clarity for the best decisions to shine forth.

Let's look at an example of false or biased information. And let's say that you were given the task to test two options in an application because of poor adoption in a critical business function of the company. Your dev team built the application from scratch, so everyone is intimately familiar with all the backend, connections, data flow, and so on.

Your testing and UX analytics work proves somewhat convincingly that Option B is the best option to go with. Yeah, but there's a problem: in order to make that work, 60 more hours of research will have to be completed, and the database will have to be updated.

The company doesn't want to do either of those. What they see is that spending the time and money to do that will put them behind the competition and cause them to lose $18M in the next quarter. So, because Option A won't take all the additional resources and can be completed in the next 10 days, they direct that course of action. In the long run, the likelihood of it all falling apart is great, but it doesn't matter; they want a quick fix.

Operational excellence relies on many factors, but one of the biggest factors is becoming UX. While there still are companies out there that don't employ UXers, those numbers are shrinking. Many giant corps have massive UX teams, and many don't even know what it is. But you're reading this book because you work for a company that does have UX and either provides it internally, externally, or both.

So, when management asks you *why* you recommend something and *how* it will be accomplished, you have to be ready for that with ultraefficient collateral. This means notes on everything. Every conversation, every person, every click, every problem or pain point, every gap, every missing piece of data... And let's not forget that we also have to report what's good and right with our world. Anything that works and gets high marks should be called out. Nothing is worse than just bad news, so if there's some good news to go along with it, use it to your advantage.

Using basic math to pivot to qualitative metrics

This is super simple, but bears a mention here to illustrate the concept. One way is to move from counts to ratios. So, instead of saying that 50 out of 90 people rated the service provided as good, you'd do the math of 50/90 = 55%. This becomes insightful when there are large disparities between ratings.

Sometimes when I've used this in the past are when I have had 'what if' scenarios, or wanted to look at what I have from different angles. Let's say I'm analyzing tickets written by the state police in Michigan.

Here's a sample table to first look at the numbers. Any time I'm doing this type of analysis, I start with a table:

Badge #	Time	Location	Speed Limit	Speed
1400	12:58:00 PM	Downtown	35	50
1345	7:35:00 AM	Midtown	25	35
1209	9:14:00 AM	Downtown	35	60
456	2:32:00 PM	Uptown	55	90
390	2:32:00 PM	Uptown	55	70
986	8:00:00 PM	Downtown	35	45
1400	2:31:00 PM	Midtown	25	40
986	9:00:00 PM	Downtown	35	55
456	4:38:00 PM	Downtown	35	50
1345	10:47:00 AM	Uptown	55	80

Table 10.2 – Ticket analysis sample table

So, the table shows several different officers, some with two tickets written, and speeding violations that vary from slight to pretty severe.

Let's do the math! Our chief wants to know what percentage of violators went more than 40% over the speed limit. First, let's get our speed limits and figure out the numbers:

- Downtown is 35

  - 40% over the speed limit is 49 mph

- Uptown is 55

  - 40% over the speed limit is 77 mph

- Midtown is 25

  - 40% over the speed limit is 35 mph

A good idea at this point would be to add a new column with these numbers to our spreadsheet:

Badge #	Time	Location	Speed Limit	40% Over	Speed
1400	12:58:00 PM	Downtown	35	49	50
1345	7:35:00 AM	Midtown	25	35	35
1209	9:14:00 AM	Downtown	35	49	60
456	2:32:00 PM	Uptown	55	77	65
390	2:32:00 PM	Uptown	55	77	70
986	8:00:00 PM	Downtown	35	49	45
1400	2:31:00 PM	Midtown	25	35	40
986	9:00:00 PM	Downtown	35	49	55
456	4:38:00 PM	Downtown	35	49	50
1345	10:47:00 AM	Uptown	55	77	80

Table 10.3 – Updated table

And a couple of visualizations of this would show the following:

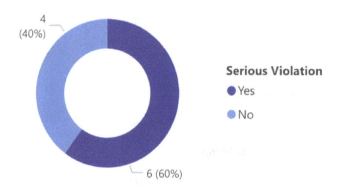

Figure 10.8 – Ratios of serious violations

You can see that this very simple donut chart with only two variables clearly shows us that the majority of the violations were considered as *serious*.

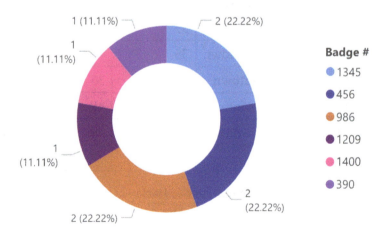

Figure 10.9 – Ratios by Badge #

So, you can see that just a little bit of math gave us new perspectives on the data to help the chief make a decision. Maybe from this, the chief will now want to determine the percentages of serious violations that occur in each location.

Summary

Well, I think you got your fill of illustrations in this chapter. I can't stress how important it is to get people thinking through visual representation of your UX analytics metrics. You got some good instruction on how to really look at, think about, and verbalize what you see in a visualization. You also learned a bit about using very basic math to give you new forms of metrics, such as ratios.

As stated, management needs direction and a strong, verified reason to move forward when making business decisions. Having an understanding of them and the world in which they live will give you UX superpowers beyond what mere mortals have, so enjoy that!

Next up in *Chapter 11*, we'll get into the intricacies of your data, and how it can make or break a company.

11
Live by Your Data, Die by Your Data

Yes, those are harsh words, I know. But it's also a business reality. Imagine a world where nobody paid attention to sales figures or customer satisfaction scores. You'd never know whether you're on track to meet your goals – and every company has sharp goals for growth, and you'd never know whether what you're doing matters to people, until you go out of business because nobody liked your product.

In the last chapter, we learned about operational measurement, and this chapter will expound on that. In my 20+ year career as a UX strategist, researcher, designer, and director, I've run across several companies that had no clue what their data is. Oh, just miss your sales goals for the quarter and see what happens! They all keep close watch on that, but not on the metrics that make it all work. And certainly not on the UX analytics – all those metrics we can gather to see whether workflows actually work and whether people are efficient at their jobs.

This chapter has five essential lessons, and we've been building up to this in all previous chapters, so there's no doubt that you'll pick this up quickly:

- Data – the beating heart of business
- Getting a sense of the scope
- Finding out what matters
- Understanding how your stakeholders want to see data
- Investigating gaps and missing data

Data is at the heart of all analytics and must be understood from several different angles. The UX analytics practitioner must become intimately familiar with all measurable aspects of their business so that they can support the goals of the management teams. Along with this comes the understanding of the taxonomy – how x relates to y, for example. Hierarchies such as a top-down of region > city > store type > storeID allow you to filter and represent differently than flat data. We'll get into that in the *Getting a sense of the scope* section. Let's dig in.

Data – the beating heart of business

No business in existence today can survive without understanding and consistently tracking its data. Watching data is how small companies grow, and it's how large companies increase efficiencies. Yes, they use it to grow as well, but large companies have so many moving parts that it's easy to overlook important data and waste money on tools and workflows that are flawed and not user-friendly.

You're learning about UX analytics in this book and the many ways in which you can gather, analyze, and either report for decision-making or make those decisions within your team. It really doesn't matter who you support within an organization; what matters is that, somehow, in some way, you and your team must be the voice of users, managers, and the data they all need to perform at their best.

We're going to explore taxonomies, dashboards, KPIs, data visualization, management views, and scope so that you understand the depth and breadth of data in the enterprise. If you think of UX analytics as a management tool, rather than just a set of practices to gather data about the usage and usability of software products, you'll get these concepts faster.

Understanding and utilizing taxonomy

In any project that my teams and I are asked to tackle, one of the first questions we ask is, "Do you have a data taxonomy?" The answer to this will give you quick clues about what you're preparing to undertake. If the person you ask looks at you like they have no idea what you're talking about – and it's more than you'd expect, I assure you – you know you have your work cut out for you. Some people know what it is; some people work with them every day. If your project is with a company that has a data department and data roles, you're in luck.

Here's a simple example of a taxonomy, after which we'll discuss the terminology and concepts:

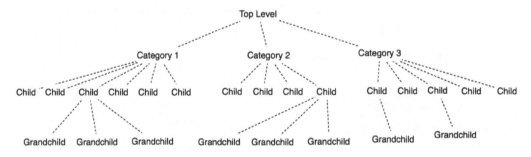

Figure 11.1 – A simple taxonomy example

Some companies talk about taxonomies in terms of parent/child relationships. You may have heard similar, or maybe different, ways of expressing it, but this is simple. In *Figure 11.1*, we have an easy example of a four-layer-deep taxonomy.

> **Important note**
>
> If you have a very deep taxonomy, this relationship terminology loses meaning quickly. Certain types of data require deep relationships, so see what you can find out first and then work backward from there.

A taxonomy is meant to portray a data structure, but it can also be a strong indicator of UI interaction patterns. If we imagine a corporate intranet site, we usually start off at the home page and then click our way down to what we need, or create shortcuts in our browser for those to save time. Yet the fact remains that the clicks take you to pages that have the information – data – you and everyone else need.

So, here now is a more detailed example of a taxonomy:

Figure 11.2 – A detailed taxonomy example

I think you'll agree that this example is only one way to create data hierarchies. While I created these relationships in this order, you may have put **Plants** higher and nested the countries underneath – and that would have been fine too. And as a UX analytics professional, you may be called on to help a company reconfigure a taxonomy.

Why? Because it affects usage, perception, intuitive navigation, and so on. A bad taxonomy can quite literally bring a company to its knees if not fixed in a way that makes sense to the users, whether those people are internal or external. And here's another tidbit for you – this sneaks into the world of information architecture, and it's not something you should shy away from. Embrace it and thrive.

I want you to be that UX pro who is full stack. Not just a designer, not just a researcher, but a UXer who has knowledge and practical experience in all phases, and being able to look at a project and know immediately how to tackle a data structure puts you at the top of the heap. You may not have an information architect to help you figure it out, but you won't need one once you master this concept and practice.

Try this exercise with your team (if you have the time, of course). Build a taxonomy from your corporate intranet, website, or the current application you're tasked with improving. I encourage it because it will help you visualize company data structure and quite possibly see gaps or misalignments that can help you in your project. And if you are on a project that's new to you and the team, you should be doing it from the start.

This should be one of your first data tasks when taking on a new project. You must be able to understand data relationships and have the ability to rearrange those that just don't work. And how do you know they don't work? Because users who you interviewed told you, and because the analytics you're gathering from the automated tools show you.

You can come to this task by way of several pain points gathered during research and analysis. This exercise of taxonomy can be led by the questions associated with the pain points. Questions such as *"Why can't the users who get into the payroll section find the form to change their deductions?"* or *"What's causing the users who want to know more about our dedication to sustainability difficulty in finding our policies?"*

Every pain point you uncover that deals with navigation problems should lead you back to the taxonomy. If you don't have one, make it right now. I created the examples in *Figure 11.1* and *Figure 11.2* in Scapple, a mind-mapping tool. There are many taxonomy tools on the market, both free and paid, so just do a search and find one that fits the needs of your company and projects.

Utilizing card sorting to fix or understand taxonomies

Card sorting is an exercise you can do in person or via a virtual session on whiteboards such as Miro or Mural. You can start with current category names (as in a closed sorting exercise), whatever those may be, and pull all the current hierarchies and click paths to start – but you won't nest everything; that's what the exercise is for. You want to see what other people think the taxonomy should be.

Look at this setup for a closed card sorting exercise, which I'll explain after:

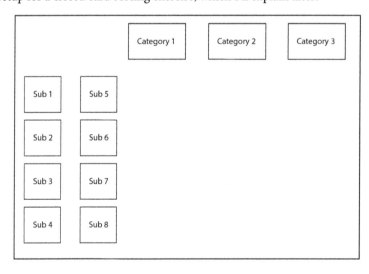

Figure 11.3 – A card sorting for a taxonomy example

Working with one person at a time in a moderated session, you'll have them move the subs under the categories they believe is the best fit. Note that these represent your current hierarchy structure, so you'd be using the correct category and sub names. If you simply pointed a participant to an online board and told them to go accomplish the exercise, that would be considered unmoderated. Here's what one new design might look like:

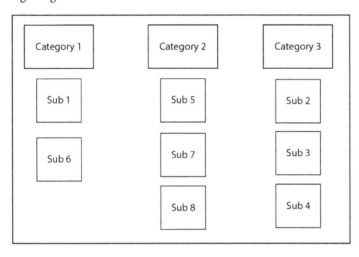

Figure 11.4 – A completed card sorting exercise example

In recent work in my company, we did this exercise with a group of managers who were responsible for website content structure. We gave them all the current categories and all the subpages, and it was enlightening to see how each arranged the cards. While most created a very similar new structure, a few felt that some subs were redundant and suggested combining. This work was all taken back to the director, and decisions were made on a new, more intuitive navigation and content structure.

Card sorting can be a very useful UX analytics tool and process. While this example is extremely simple, larger, more complex content structures will require several sessions with managers, content writers, and most certainly users of the application – who all will give valuable input and ideas toward solutions.

Another way to look at a taxonomy

Figure 11.1 and *Figure 11.2* are great visual ways to show taxonomies, but they aren't the only way. If you think about folder structures with files in them, that's a form of taxonomy. And I created the figures to conserve page space, but they can just as easily be created and read left to right.

When needing something even simpler, just to work through quick ideas with the team, these hierarchies can be written like this (which you might already be familiar with):

Company > Departments > Jobs > People

Alternatively, to show a simple click path to illustrate a bit deeper, it could be the following:

Company > Financials > Accounts Payable > Invoices

No matter how you choose to do this, the end result should look something like *Figure 11.2*. It most likely will be wider and deeper, and if you have years' worth of content, it will get extremely wide.

Getting a sense of the scope

I think you'll agree that, now that we've dipped our toes into the taxonomy pool, we're starting to understand the scope. I'll reiterate that every analytics project will involve people, data, systems and tools, and corporate needs. Since our focus in this chapter is data – in a deep dive – then to understand the scope, we need to look at data and people, data and systems, data and tools, and data and corporate needs.

If your project is internal and your company is 10 years old and has been collecting and storing data for six of those years, you may have a lot of historical data. This is the ideal situation but, unfortunately, is still not a reality for most companies across the globe.

Scoping breadth and depth

How wide is your taxonomy, and how deep does it go? Wide and shallow taxonomies are considered to be flat. This means that there are many categories, and very few subcategories nested underneath them. This can be caused by many things, including being too granular at the higher levels. I'm going to stick with a retail example again because it's easy for us all to visualize – we all shop, online or in a physical store, but we all do it.

A flat taxonomy might have same-level categories for all types of canned foods:

- Canned foods:

  - Canned peaches

  - Canned green beans

  - Canned pasta sauce

  - Canned tuna

  - Canned meat spread

  - Canned cheese spread

  - Canned tomatoes

  - Canned coffee

A more intuitive and deeper version might have higher-level categories:

- Canned fruits:
  - Canned peaches
  - Canned prunes
- Canned vegetables:
  - Canned green beans
  - Canned sauerkraut
- Canned meats:
  - Canned meat spread
  - Canned goose liver
- Canned seafood:
  - Canned sardines
  - Canned oysters
  - Canned caviar

A deeper taxonomy that considers categorization levels helps create a more user-friendly and intuitive application interface. Your analytics practice and projects will either prove or disprove the design decisions that were made in the past and implemented. If you find that people abandon a process, an online shopping cart, or sending emails to the help desk often because they can't find what they want, you must look at navigation structures and figure out how to fix them.

Finding database information

This is another area you need an understanding of. Not how to make and work with one, but how it's accessed through applications and how much is in it. How big is the database that serves Application Mightymite? How much data do we have on manufacturing throughput over the past five years? Where is all the data on the number of passengers who fly transatlantic routes regularly?

Whatever the data sources are, and whatever purposes it all serves, you need to have knowledge of this. It certainly helps if you have a data team to work with, but this won't always be the case. If you do, you want to get to know them so that you have an ally when you need to get more information than you have direct access to.

Even better is if you have a data science team, or even just one data scientist to work with. They can write the algorithms that get you what you need if the current numbers don't give you the entire picture. Find them and make friends – they can help you immensely with challenging data problems.

Find out what kind of databases your company has – Oracle, SQL, MongoDB. Relational, object-oriented, cloud, distributed. Again, I'm not saying you need an intimate knowledge of any of these, but knowing what you have, and a basic understanding of how they work, will be to your advantage when researching problems and pitching possible solutions. If you'd like to take this a step further, look online for courses on data modeling so that you can have a deeper understanding and can speak the language when talking to your data people.

Finding out what matters

What matters is a combination of what a corporation says matters and what strategic and task-oriented employees say matters. While there are goals and accomplishments that the enterprise says must be realized, there are also real-world considerations from the ground up. *Figure 11.5* puts this into perspective.

Figure 11.5 – The sweet spot of what matters

Organization: "We need this and we need it now!"

Individual(s), the ones you interviewed, and the ones from whom your automated analytics gave you metrics: "Yeah, okay, sure. Just fix this *XYZ* software so that I don't have to spend 45% of my workday trying to find what I need, in the six different places I have to look, and then combine all that together with Excel and Word."

When stakeholders understand what they're up against, because you've done the right kind of UX analytics work and proved the slight flaws in the top-down plan, that plan can be adjusted to fit the real conditions of a company and get to an alternative, but equally acceptable, outcome – perhaps for phase one, with the second phase coming shortly after the internal problems are solved.

Companies that excel while others go through round after round of layoffs conduct business this way. They set goals, and then test those goals against the information and data gathered by UX teams. They don't make snap decisions that aren't founded on trustable information, metrics/data, and relevant intelligence pertinent to the mission.

Where you have more agreement than disagreement on what a company says and what the task workers say, you can certainly move faster at the starting stages because you know that not many alternatives need to be presented and a big pivot isn't necessary.

There will be times when the obvious solution is also the easiest, but yeah, not always. This is where a feasibility diagram comes in handy. It's a tool and mechanism that allows an entire cross-functional team of researchers, designers, developers, engineers, and managers to work together and figure out a workable path forward. This is a simple four-quadrant diagram where the team will use sticky notes in design thinking work to assess and vote on approaches. Here's one:

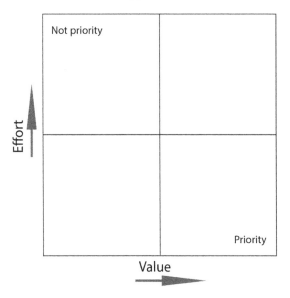

Figure 11.6 – A feasibility diagram example

Based on UX analytics findings that start to lead a path to a solution, the most pressing pain points and challenges, the ones that were voted on by most people as being time wasters (and thus money wasters), will be put to this test. They'll be written out on sticky notes and placed on this chart, based on the effort it will take to accomplish and the value it will provide to users and customers.

The bottom-right corner of the preceding figure will be most likely to be incorporated, due to the low effort to accomplish it and the high value it delivers, whereas the exact opposite is true for the upper-left quadrant. Those involve high effort and low value, so they will either get discarded or moved to another phase of the project. You may also find out that they're dependent on work that's in the bottom right, which means that when that work is accomplished, those can be moved down to less effort work, and although the value might not be very high, they can be incorporated to round out the project.

Understanding how your stakeholders want to see data

Think back to *Chapter 8* where we introduced the Stakeholder interview template. In question 7, we asked for KPI data. And this is the part mentioned there about getting more clarity, so here we go!

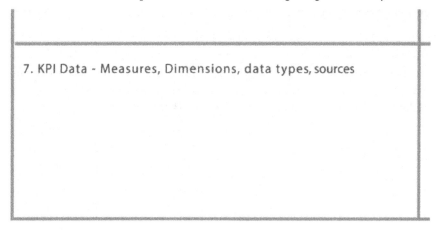

7. KPI Data - Measures, Dimensions, data types, sources

Figure 11.7 – The stakeholder KPI question

Let's imagine that some of the KPIs they mentioned during the interview aren't simple numbers from a column in a spreadsheet or database somewhere. They could be derived or calculated from other data and are not easy to find. And remember too that the numbers that stakeholders want to see are often higher-level metrics that are a few levels above what our users have in their spreadsheets.

I've found in my work that sometimes the stakeholder isn't aware of where KPIs come from. If during the interview you probe a bit further into KPIs and discover their origins, you're okay for the next steps, but if not, you'll have to backtrack a bit, which is okay because our stakeholder is our ally in this project.

It also may be the case that the stakeholder needs new information. You may get to the KPI portion of the interview and be told that because of something new happening in the company, or new goals, or a new product that's never hit the market before, they will now need to track a few new numbers.

And to ensure that nothing is being overlooked, it should also be your job to look at the numbers in a way where you can derive new meaning. So, once you and your team have looked at all the KPIs and have an understanding of what they are and where they come from, you can now do *what-if* scenarios.

What if we divided this number by that number? What if we found a way to show this set of numbers in a better way? What if we could prove that by looking at this KPI from a different angle, it would enhance the analytics and allow for more accurate decision making?

The point here is to never take anything at face value, no matter the source. UX analytics requires that of you. It requires that you're constantly asking questions and looking at challenges from fresh perspectives. This is why a cross-functional team is essential. Each member of the team thinks in different ways, and together you all work to deliver the best answers possible. And never forget that those answers will often lead to new questions.

Visualizing the KPIs for quick decision-making

We've seen examples of data visualization, and we're going to go wider now to what are commonly referred to as *dashboards*. These are interfaces that contain several different visualizations of the KPIs. In my workshops, I take a company through exercises toward the end where we all create dashboards individually, then work together to pull the best ideas out of all of them, and create a mashup that fulfills the requirements for the stakeholder's needs.

Here's a wireframe that provides enough structure to allow team members to think through the design, but not lead them toward anything biased in a particular direction.

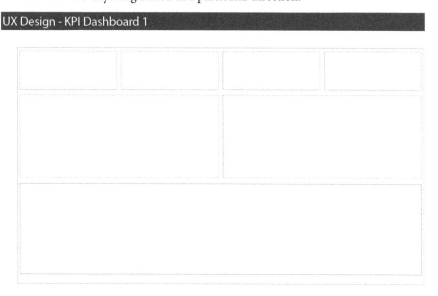

Figure 11.8 – A simple KPI dashboard wireframe template

We saw KPIs in *Chapter 3*, so these would most likely be what goes into the smaller sections at the top of the previous figure. You want to ask the stakeholder what these are. If there are more than four, make room for them. This is just a start to get people thinking about the interface, so learn to be flexible, and teach that to the team.

KPIs can be shown on a manager-level dashboard, but make them BIG:

Sales This Quarter	Losses This Quarter
$42 M	$2.4 M

Figure 11.9 – A large KPI numbers example

And to move us along quickly here and tie this part up, here's an example of a simple KPI dashboard for a stakeholder.

Figure 11.10 – A KPI dashboard example

The bar charts, donut chart, and scatterplot with bubble sizing all support the numbers in the KPI cards. Also, note that each of the four supporting sections is tabbed so that other types of similar supporting visuals can be included. This is a quick and simple way for your team to create designs to take to the stakeholder – who may have even worked on these – and determine in an iterative fashion what would get greenlighted and given to the development teams to produce.

Ultimately, the final decisions will be made with the stakeholder so that your team has delivered the right tool for the job at hand. *Figure 11.10* is just a simple example of what's possible to deliver from your analytics efforts, so stay open and flexible when doing this work. Don't feel locked in until it's time for management to make a decision.

Investigating gaps and missing data

Now, you need to roll your sleeves up and get busy. When you start to gather data from what people work with, you start what I like to call *data archaeology*. You're digging and poking and prodding a spreadsheet most likely, and you're finding structures and categories and holes and gaps. This is interview process territory when you're asking users to provide you with samples. You can also get this if you ask for and are given user-level access to the application or systems they're using.

The best way to understand the process here is by example, and I'm going to walk you through a simple but effective one. This practice encompasses the two primary facets of UX analytics – research and analytics investigation. You do the research to uncover the data needs of the users, and then you start investigating what you're given to find answers to the pains and challenges. These are your clues, and you must follow them through.

Finding and fixing data problems

For the purposes of space in this book, the sample will be small, but it will help to illustrate data gap problems and lead you to understand how many types of questions you can ask to get to the root cause of a problem.

Here's a sample from a spreadsheet that's simply a set of made-up data:

Product ID	Date	Product	Amount	Sales Person
10	5/4/22	Beanie	$ 17.00	100
10	5/4/22	Beanie	$ 17.00	
10	5/4/22	Beanie	$ 17.00	103
10	5/5/22	Beanie	$ 17.00	110
10	5/5/22	Beanie	$ 17.00	100
19	5/3/22	Black Belt	$ 6.00	
14	5/5/22	Black Tshirt	$ 16.00	102
14	5/5/22	Black Tshirt	$ 16.00	102
13	5/4/22	Brown Belt	$ 12.00	105
9	5/3/22	Gloves	$ 26.00	
8	5/3/22	Scarf	$ 20.00	103
3	5/3/22	Socks	$ 4.00	103
3	5/3/22	Socks	$ 4.00	103
3	5/4/22	Socks	$ 4.00	
5	5/5/22	Sunglasses	$ 14.00	101
7	5/3/22	Tie	$ 10.00	101
7	5/4/22	Tie	$ 10.00	
2	5/5/22	White Sneakers	$ 30.00	107
17	5/4/22	White Tshirt	$ 14.00	107
16	5/4/22	Yellow Tshirt	$ 14.00	107

Figure 11.11 – A simple data example

Since this is a short sample of data, it's easy to see that in the **Sales Person** column, there are cells with no data. If this spreadsheet had 1,000 rows, you'd certainly find more empty cells. Look at this quick representation I did in **Tableau Public** to visualize the gaps:

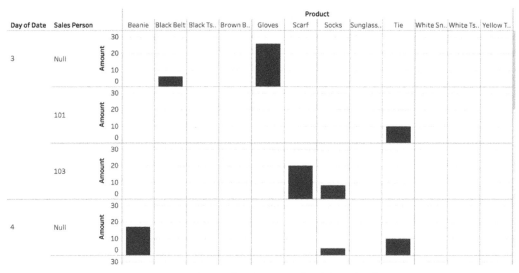

Figure 11.12 – A visualization example of the sales data

The two instances of **Null** in the **Sales Person** column mean that there's missing data. If we then associate the products sold – **Beanie**, **Black Belt**, **Gloves**, **Socks**, and **Tie**, we can start to see whether there's any type of pattern. One **Null** is on day **3**, and the other is on day **4**. **Gloves** was the highest sale on day **3**, and if these salespeople are on commission, somebody is going to be sorry on payday.

Your UX analytics work has just proven that there's a problem somewhere. Maybe someone even alluded to this in an interview: *"Sometimes my commissions don't add up and I have to fight with payroll to get what I'm owed."* So, what's next? What do you do with the information you've gotten and the problem you've proved exists? You ask more questions.

Determining where the problem is

In this scenario, the problem is that data is missing for sales of certain items on random days. This leads to the next problem of salespeople missing commissions. You put your analytics hat back on and you get back into data archaeology mode. Here are some questions you might ask to start sleuthing:

- Why is this data missing? (Of course, this is the first question.)

- Is it a point-of-sale hardware problem? Maybe when the touchscreen gets dirty toward the end of the day, it doesn't register certain inputs:

  - Can we further break down the data so that we can see the time of day and determine whether this could be the reason?

- Is it a user error problem? Maybe these are newer employees who don't understand all they need to do to enter the sales correctly:

  - If we find this to be true, should we take a look at the onboarding and training documentation to see whether we need to rewrite something?

  - If we find this to be false and the employees have been around for a while, how can we be certain that they're following a procedure?

- Is it a database problem? Maybe there's a bug and this is proving it:

  - Who do we show this problem to?

  - What will it take for them to check – what do they need from us? More data perhaps?

- Is it a data type problem? While this seems unlikely, it's possible:

  - Is there something about the **stock-keeping unit** (**SKU**) number for that item that doesn't work with data rules, schema, or type?

  - Does it have something to do with that employee's ID number?

    - Maybe they aren't entering it correctly.

- Is it an application interface problem? Maybe something is unintuitive, or maybe touch buttons are placed too close together:

  - Let's talk to the store manager and have them talk to the employee to investigate for us.

  - Let's find out how many of the other employees experience a difficult thing to do with the interface. If we find that others have this problem occasionally, we may have a bigger clue.

- Is it a combination of problem types?

  - Could it be newer employees who've taken the training but find the interface difficult to use?

  - Could it be a combination of a dirty screen and poorly placed touch buttons?

  - Could it be $x + y + z$?

So, you start asking these questions to see where they lead. You speak to the people responsible for each of the areas, and you start to narrow down the possibilities so that you can report on them and start moving toward a solution. This also will help tie up loose ends if they exist. We spoke about "trustable" data in a previous chapter, and this is where you want to be diligent and do what is needed to provide that level of confidence in the data presented and made available to your users.

Summary

Live by your data, die by your data. You understand that at a completely different level now, don't you? You're getting a deeper appreciation for what data means to a company, how to find it, work with it, and present it for decision-making.

You learned how to work with stakeholder data and how to dig in with data archeology to get to the root of a problem by asking questions, and you learned more about data visualization and its impact on enterprise-level thinking. You learned the effective use of feasibility diagrams to assess problems by comparing value and effort, which allows for a development roadmap to be more accurate and indicative of the work and investments needed in money and labor to accomplish new application enhancements.

Next up, in *Chapter 12*, we'll get an overview of UX analytics tools and which are best for your particular situation. There are many to choose from, and this book isn't the place to dive into all of them in depth, but you will be able to make decisions on the best fit for your needs.

12
The Tools It Takes to Gather and Consolidate Information

When it comes to understanding user behavior and improving the user experience of digital products, the collection and analysis of data and information collected during your research phases are key. And it's true that sometimes, gathering and consolidating data from various sources can be a challenging and time-consuming process.

This is where the right tools come in, making it easier to collect, analyze, and visualize data in a consistent and reliable manner. In this chapter, we'll explore the various tools needed to effectively gather and consolidate UX analytics data, and how these tools can help your stakeholders and organization make informed decisions based on accurate insights.

Through various examples, we'll provide you with the best direction for your specific projects and needs, and help you gain an understanding of the different tool types available by covering the following topics in detail:

- A brief overview of UX analytics tools
- How to determine what is best for your needs
- How to quantify findings for reports

A brief overview of UX analytics tools

UX analytics tools are designed to gather and analyze data related to user experience. They help UX teams within companies understand how users interact with their digital products, which can include your websites, mobile apps, and software applications used either internally or externally.

There are various types of UX analytics tools, including tools for gathering data through surveys, interviews, user testing, heatmaps, click tracking, and A/B testing. Each tool has its own set of advantages, disadvantages, and best practices, and choosing the right tool depends on the specific goals of your research project.

Once data has been collected, UX analytics tools are also used to analyze and visualize the data. Popular tools for analyzing this data include **Google Analytics**, **Mixpanel**, **Adobe Analytics**, **Hotjar**, **Crazy Egg**, and **UsabilityHub**. These tools can help you and your team identify patterns and trends in user behavior, understand how users navigate through your digital products, and identify areas for improvement.

Choosing the right UX analytics tools is critical for effective data collection and analysis because you must ensure the highest quality reports are delivered to those needing to make decisions. It's important to consider factors such as the goals of the research project, the target audience, the complexity of the digital product being analyzed, and the project's budget.

Overall, UX analytics tools play a crucial role in helping teams within organizations create digital products that are user-friendly, intuitive, and effective in meeting the needs of their target audience.

Evaluating and selecting tools

When it comes to selecting tools for gathering and consolidating UX analytics data, there are several factors to consider. The most important consideration is the specific goals of the research project, as different tools may be better for evaluating certain types of research questions or metrics gathered. Other factors to consider include the size of the user base, the complexity of the product, and the level of detail required in the data analysis.

In order to evaluate the various tools available, it can be helpful to create a list of features and functionalities that are important to the project goals. This can include things such as data collection methods, data visualization capabilities, and ease of use. Additionally, it can be helpful to read reviews and recommendations from other users to get a sense of the pros and cons of different tools. Since this is the age of user reviews and star ratings, be sure to include it in your evaluations.

Ultimately, selecting the right toolset for a research project requires careful consideration and evaluation of the available options. It's important to choose tools that are well suited to the specific goals of the project, and that offer the right balance of features, functionality, and ease of use. By taking the time to carefully evaluate the available options and make informed decisions, you can be sure that you're collecting accurate and actionable insights that will drive meaningful improvements in the user experience of your products.

Next, we will explore the tools mentioned in the opening part of this chapter in a bit more detail.

Google Analytics

Google Analytics offers a wide range of features, including tracking of e-commerce transactions, user behavior analysis, custom reporting, and integration with other Google products such as **Google Ads** and **Google Search Console**. It also has the ability to track user behavior across multiple devices and platforms, making it a valuable tool for businesses that have a strong online presence. Here are two key points about Google Analytics:

- It uses a tracking code that's placed on each page of a website to collect data. This data is then processed and displayed in the Google Analytics dashboard, which provides an overview of key metrics, as well as detailed reports and insights into user behavior.

- Overall, it's a powerful tool that can provide you with valuable insights into how your website is performing, and help you make data-driven decisions to improve your company's online presence.

Mixpanel

This is a web and mobile analytics platform that allows you and your team to track user behavior and engagement across your digital products. It offers a range of powerful features including event tracking, funnels, cohorts, A/B testing, and segmentation, the benefits of which are as follows:

- Event tracking is one of the most important features of Mixpanel. It allows businesses to track specific actions taken by users within their digital products, such as button clicks and form submissions.

- Funnels are another important feature of Mixpanel. They allow businesses to track the steps users take to complete specific actions, such as making a purchase or signing up for a newsletter.

  - They can also help you identify points of friction in the user journey and optimize your products for better user engagement.

- Cohorts allow businesses to group users based on shared characteristics or behaviors, such as first-time users or users who have made a purchase within the last 30 days.

  - This can help you identify trends and patterns in user behavior and tailor your products and marketing efforts accordingly.

  - If you're unfamiliar with cohorts, a good example might be age range. For example, you can bucket new registrations to your app by those aged between 15 and 30, and then those over 30. This way, you can track their behavior over a period of time to see what they do in your app.

- A/B testing is another feature of Mixpanel. Testing different variations of your products and measuring their impact on user behavior and engagement is a powerful way to determine whether you need to redesign or drop features, or add new ones in.

- Finally, segmentation is a powerful feature of Mixpanel. This allows you to group users based on a wide range of criteria, such as demographics, behavior, and location.

  - By knowing this, you and your team can understand how different user segments interact with your products so they can be designed accordingly.

  - This is similar to a cohort but a bit broader in scope, so weigh your needs and then decide which is the best way to go.

Adobe Analytics

This is a web analytics and reporting tool that helps businesses to track and analyze user behavior across various digital channels, including websites, mobile apps, and social media platforms. It provides you with advanced segmentation and reporting features, and offers the ability to create custom dashboards, reports, and alerts to monitor key metrics and KPIs.

There's a wide range of features to make use of, so here's a summary for you:

- **Real-time data tracking**: Adobe Analytics can track user behavior in real time, allowing you to quickly identify trends and patterns.

- **Custom dashboards**: You and your team can create views and provide access to dashboards that display key metrics and KPIs, and monitor changes in real time.

- **Advanced segmentation**: Adobe Analytics allows you to segment your data based on a wide range of criteria, including demographics, location, and behavior. This is similar to segmentation and cohorts in Mixpanel.

- **Multichannel attribution**: Adobe Analytics provides insights into how users interact with your brand across various channels, including social media, email, and mobile apps.

- **Machine learning capabilities**: Adobe Analytics incorporates machine learning algorithms that help you identify and predict trends and patterns in your analytics data. This is a powerful feature as machine learning and AI become more prevalent in tools and systems.

Hotjar

Hotjar is a behavior analytics tool that tracks website user behavior through features such as heatmaps, session recordings, and feedback polls. It helps companies understand how users are interacting with their websites and identify areas for improvement. The benefits of these features are as follows:

- Heatmaps are among the most popular features of Hotjar, as they allow website and web-based application owners to see where users are clicking, tapping, and scrolling on their website. By seeing this data in the form of a heatmap, you can identify which areas of your digital product are getting the most attention and which areas are being overlooked.

- Session recordings are another useful feature of Hotjar, as they allow you and your team to view recordings of user sessions. This can be helpful in understanding how users are navigating through the website or application, and where they may be encountering issues or obstacles.

- Finally, surveys and feedback polls are also available in Hotjar, allowing you to gather feedback directly from your users. This can be helpful in understanding why users are behaving in a certain way on the website, or in identifying areas for improvement. (Remember that you learned a lot about surveys in *Chapter 9*.)

Crazy Egg

This is a website optimization tool that uses heatmaps, scroll maps, and click reports to track user behavior on a website. It provides insights into user behavior, such as where users are clicking and scrolling, and helps companies optimize their websites for better engagement and conversion rates. Further benefits of Crazy Egg are as follows:

- It also offers a range of other features such as A/B testing, session recordings, and user surveys, similar to the other tools mentioned.

- One of the benefits of Crazy Egg is its ease of use, with a user-friendly interface that allows even non-technical users to quickly set up and run tests. It also offers integrations with other popular tools such as Google Analytics and **Shopify**.

- The scroll map function mentioned earlier is a popular and effective feature. It shows how far users scroll down a page before leaving. This can help you identify which sections of a page are most engaging and where users might be losing interest.

UsabilityHub

This is a user research platform that offers tools such as remote usability testing, click testing, and five second tests. It helps companies gather feedback and insights from users about the usability and effectiveness of their digital products. Its suite of tools includes the following:

- **Five second test**: This allows you to quickly test the effectiveness of a landing page or design by showing it to users for five seconds and then asking them a set of questions. With quick flashes like this, the answers you get can be very powerful indicators of your interface.

- **Click test**: This lets you test how easy it is for users to find specific elements on a web page by asking them to click on certain areas and tracking their mouse movements.

- **Nav flow**: This tests the effectiveness of your website's navigation by asking users to complete a set of tasks and tracking their movements through the site.

- **Question test**: This allows you to test the usability of your website or app by asking users a set of questions about their experience.

- **Preference test**: This helps you understand user preferences by asking them to choose between two or more design options.

UsabilityHub's tools are designed to be easy to use and require no technical expertise, making them accessible to a wide range of users. The platform also provides detailed analytics and reports to help you make informed decisions about your designs and user experience. Overall, UsabilityHub is a great option for businesses and organizations looking to gather quick and reliable feedback on their digital products.

So, to wrap this section up, gathering and consolidating information through the right tools is crucial to gaining insights into user behavior and making data-driven decisions. From Google Analytics and Mixpanel to Adobe Analytics, Hotjar, Crazy Egg, and UsabilityHub, there are many powerful tools available to collect, analyze, and visualize data. Selecting the right toolset involves considering the specific goals of the research project, evaluating the tools, and making recommendations for the best fit. By using the right tools for data collection and analysis, you and your team can optimize your digital products and ultimately improve the experience for all user types and personas.

Determining what is best for your needs

If we think back to all we've learned in the previous chapters, we can see that we've already done the work needed to understand the problem. We learned about user interviews, surveys, data, and metrics that have meaning to our company, and we've established a strong foundation for all the work we must accomplish to realize our company's goals.

So, with that in mind, there are a few key considerations that can help you determine which automated UX analytics tools are best for you and your project's needs.

First, it's important to clearly define the goals and objectives of the research or analysis that the tools will be used for – which, at this point, you know is crucial. And from that work, you've identified the types of data that need to be collected and the specific features and capabilities that the tools should have.

Defining the goals and objectives of research or analysis is a critical step in selecting the appropriate toolset. This helps ensure that the data collected is relevant, reliable, and actionable – with actionable being the key word. No matter how much work you've done, if what you produce can't be acted upon by someone whose job it is to carry on the work, it won't get done.

Remember that you must clearly identify the research questions or hypotheses and determine what data needs to be collected to answer them. And keep in mind that the goals and objectives of the research and analysis may vary depending on the stage of the product development cycle or the business's overall strategy.

For example, if the goal is to identify usability issues in an application, a tool such as Hotjar or UsabilityHub may be appropriate. If the goal is to track user behavior and engagement, a tool such as Mixpanel or Google Analytics may be a better fit. By clearly defining the research objectives, you and your team, and ultimately your stakeholders, can make informed decisions about which tools to use and ensure that the data collected provides meaningful insights.

Companies should also consider their budget and resources when selecting tools. Some tools may be more expensive than others, and some may require more technical expertise or training to use effectively.

So, when considering the budget for choosing the right tool for a project, it's important to remember that cost isn't the only factor to consider. While some tools may have higher upfront costs, they may also provide more robust features and capabilities that can save time and money in the long run. You'll also find that some tools have lower initial costs but require ongoing maintenance and updates that can add up over time, so take this into consideration.

One approach to managing budgets for tools is to prioritize needs and allocate resources accordingly. This can involve identifying the most important features or capabilities needed for a project and focusing on tools that provide those features at a reasonable cost. It may also involve setting a budget for tools upfront and sticking to it, while being open to adjusting the budget if needed based on new information or unexpected costs.

Another approach is to consider the **total cost of ownership** (**TCO**) of a tool, which includes not just the initial cost but also ongoing costs such as maintenance, training, and support. This can help ensure that you understand the true cost of a tool upfront, rather than just focus on the initial price tag.

The training part should never be ignored. If you have a team that's proficient in many types of tools and has possibly used the tool you're considering in a previous project, this might not be an issue, but don't discount the need for ensuring that training will be minimal or that you have staff who can get up to speed quickly. If training will take a while, be sure to build this into your project deliverables and timelines. You don't want to promise something you can't deliver on time.

Ultimately, the key is to balance the needs of the project with the available budget, and to be open to exploring different options and approaches to achieve the desired outcomes.

It can be helpful to read reviews and compare different tools to see which ones have the features and capabilities that align with your needs. It's also a good idea to consider the support and customer service provided by the tool vendors, as well as any additional resources or training that may be available.

When reading reviews and studying what others have said about a product, it's important to look beyond just the star ratings or overall sentiment. Here are a few additional tips:

- Look for reviews that are specific to your needs and relevant to your use case. These will be more informative and help you determine whether a particular tool is a good fit.

- Pay attention to common criticisms. If you notice that many reviewers are mentioning the same issues, limitations, or even usability problems, it's a sign that those issues may be significant and worth paying attention to.

- Always consider the source of the reviews. Reviews from reputable sources such as industry publications, trusted influencers, or peers in your network can carry more weight than anonymous reviews or those from sources you don't recognize. No matter how good a tool might be, there will always be negative reviews, so look for a ratio. Anything over 70% positive may be worth your time to investigate further.

- However, don't discount tools with negative reviews outright. Even if a tool has some negative reviews, it may still be worth considering whether those issues do or don't impact your specific use case or whether the tool has other features or benefits that outweigh the negatives.

- Always look for reviews that discuss the specific features you're interested in. Reviews that focus on the features you care about will give you more insight into whether a tool can meet your needs.

By following these tips, you can get a better sense of how a tool performs in real-world scenarios and make a more informed decision about whether it's the right fit for your project and budget. Look at the long term, as well. You don't want to be spending company money on a tool that only fits a specific, one-off use case. Think about your work overall, and the future work you'll be tasked with completing.

Finally, it can be useful to conduct a trial or pilot of the tools before committing to a long-term subscription or purchase. This can help ensure that the tools are a good fit and that they are able to provide the desired insights and value.

A trial or pilot involves testing the tool's features and capabilities to ensure that it meets the project's goals and objectives, as well as assessing its ease of use, data quality, and compatibility with other systems or tools.

This assessment period can also help identify any limitations or issues with the tool that may not have been apparent from reading reviews or studying its documentation. This allows you and your company to make an informed decision about whether to invest in the tool for the long term or seek out alternative options.

When conducting a trial or pilot, it's important to define clear criteria for success and to collect feedback from all relevant stakeholders. This feedback can help inform the decision-making process and ensure that the chosen tool is the best fit for the project's needs.

It's also important to consider the duration and scope of the trial or pilot. A short-term trial may not provide enough data to fully evaluate the tool's effectiveness, while a pilot that is too large in scope may be resource-intensive and costly. Striking a balance between these factors is key to conducting an effective trial or pilot.

Quantifying findings for reports

Whether or not you've purchased a tool, you have to show your research findings and analytics metrics in a way that people will understand. By quantifying your research, you're putting the information into a form that matters most to management – cost and **return on investment** (**ROI**).

Quantifying research findings and analytics metrics is an essential step in making data-driven decisions and demonstrating the value of UX research and analytics to stakeholders and decision-makers. Quantification involves using numerical and statistical data to measure and compare user behavior and experience, and presenting this data in a way that is easy to understand and relevant to business goals.

This approach can help to illustrate the impact of UX design changes or other improvements to digital products, and to make a compelling case for investing in further UX research and analytics. Some common metrics used to quantify UX research findings include conversion rates, bounce rates, time on site, and click-through rates. These metrics can provide insight into user engagement, behavior, and preferences, and can be used to track progress toward business objectives and goals. In addition, it's important to consider the audience when presenting research findings and metrics, and to use visualizations and language that is easily understood by non-technical stakeholders.

When presenting your findings and analytics metrics, it's essential to remember that the intended audience may not have the same level of technical expertise or domain knowledge as the research team. Therefore, it's critical to communicate in a language that the audience can understand. Using technical jargon or highly specialized terminology can be confusing and may even cause stakeholders to question the validity of the findings.

Therefore, it's essential to present data in a way that is accessible to non-technical stakeholders, including visualizations such as graphs, charts, and tables that are easy to interpret. In addition, using plain language to describe the findings and metrics can help ensure that stakeholders can understand and act on the insights provided. By communicating research findings in a way that is accessible to all stakeholders, teams can facilitate more informed decision-making and increase the impact of their research.

I think a few examples are called for here, and these will help solidify the lessons you learned in *Chapter 10* and *Chapter 11*. To get working with some numbers you may encounter, let's create a simple scenario.

Quantitative data scenario for an application

Let's assume that you and your team are charged with researching the viability of adding a new feature to your business-to-business customer order portal. This is a web-based application that business customers are set up with so that they can place self-service orders for the products they buy from your company.

Research has shown that customers want a way to use an existing or past order to copy for reorders. Currently, every new order has to be started from scratch and there's no automated way for them to place a repeat order quickly.

Further research shows that because of this, your customers are ordering more and more of their products from your competitors who have an easier system for them to use. They admit that the quality of the competitors isn't the same, but the ease of use of their ordering platform causes them to choose them.

Here's what you need to establish (and these are just off-the-cuff numbers to illustrate):

- The amount of time it will take your development and engineering teams to add this new feature to the application:

  - What most companies do here is figure out the exact roles it will take and an average cost per hour for those people:

    - `$135/hr`

  - They then estimate the total hours it will take to complete the enhancement and the cost that will incur:

    - `200 hrs * 135 = $27,000`

  - They would also want to figure in the time and effort it took the UX research team to complete the work that proves the need for the enhancement:

    - `45 hrs`

- That time translated into dollars for the total cost of the work:

  - `$95/hr * 45 hrs = $4275 + $27,000 = $31,275`

- The amount of time it takes your customers on average to complete a new order:

  - `45 minutes`

- The amount of money your customer is losing because of the difficult ordering process:

  - `45 * 3 people * $35/hr * 22 days per month (avg) = $103,950 per month`

- The amount of money your company is losing per month when customers order from a competing company:

  - `$168,000`

- The amount of revenue your company can regain by providing customers with an easy-to-use system:

  - `$2,016,000`

- The amount of new business your company can gain per year with the newer, easier-to-use ordering application:

  - `$4,500,000`

- The estimated return on investment and timeframe:

  - `$6,516,000 - $31,275 = $6,484,725` over a three-year period
  - The `$31,275` will be recouped in 1 to 2 months

So, it's quite obvious that the work will be paid for very quickly, and you have a strong case for regaining the lost customers by showing them how much these enhancements will save them in both the short and long term. When you use your research in this way, first establishing sentiment and finding the core problems customers are having, then getting to the dollars and cents of the situation, you will provide a very strong case to the business stakeholders and managers and, just as importantly, to the customers who have abandoned ship because of their pains in working with your company.

Keep in mind, too, that the given numbers are just examples of estimates of time. Everything varies and there will be times when you can come in under budget, and times when pivots occur and you have to start over. You could be close to the finish line and disaster strikes, so be sure to spell it out if and when you pad the budget for those possibilities.

Summary

In this chapter, you learned about a variety of UX analytics automation tools and what their capabilities are. You know that these tools can be a part of your arsenal but be wary of using only them – your interview skills are getting stronger and your methods are getting tighter, so it's the combination of all the methods that gets you the most return for your time and efforts.

You saw examples of quantitative metrics that show losses associated with bad application UX and how to report them in your readout. Always remember the importance of talking to business people in a language they understand money. Any time you can show losses compared to an investment to recoup those losses and gain new revenue, do it.

Next up, we jump to *Part 4*, beginning with *Chapter 13*, in which we'll explore digital transformation and the necessity of ensuring that an organization is prepared for the changes to come, especially if it's struggled with mostly manual processes in the past.

Part 4:
Strategy

This part explores the realm of strategy in the context of UX, highlighting key aspects and practices that drive successful digital transformations and effective content management. In today's rapidly evolving digital landscape, organizations are embracing digital transformation as an enterprise practice to eliminate time-wasting manual processes and optimize efficiency through the integration of digital tools and applications. This part begins by diving into the concept of digital transformation, showcasing how it can revolutionize businesses and enhance user experiences.

In the realm of UX, content serves as the visible face of a company to users. Just as data and research form the foundational structures of UX analytics, content is the structure that users interact with directly. Therefore, capturing, storing, and ensuring seamless access to content are crucial aspects of a digital business strategy. This part explores the importance of content management and its role in shaping the user experience.

This part equips you with the knowledge and insights necessary to develop effective strategies for digital transformation and content management. By understanding the principles and practices outlined in these chapters, you will be well prepared to navigate the complex landscape of strategy and leverage it to drive meaningful business outcomes and deliver exceptional user experiences.

This part has the following chapters:

- *Chapter 13, Digital Transformation – An Enterprise Use of UX Analytics*
- *Chapter 14, Content Strategy and Analysis – An Analytics-Based User-Centric Approach*

13

Digital Transformation – An Enterprise Use of UX Analytics

Digital transformation is the enterprise practice of removing time-wasting manual processes by streamlining those processes with digital tools, applications, and systems. By automating manual processes and leveraging data-driven insights, companies can gain a competitive edge in their respective markets, while enhancing the overall user and customer experience.

In this chapter, we'll explore the role of UX analytics in digital transformation and how it can help your company achieve its business objectives. We'll cover the following topics:

- The need for digital transformation
- How UX analytics supports the plan
- The leadership team and its influence

In each section, we'll continue with everything you've learned up to this point and show how and why what you do as a UX analytics specialist supports company agendas and, if there's currently nothing happening with digital transformation, how you, your team, and your stakeholders can spearhead this important management strategy.

The need for digital transformation

Our world never rests in the 21st century. AI is transforming the way we work, play, and relax. In companies from mega-corporations that make billions a year to start-ups and mom-and-pop shops, there's a thread of technology in each and every one.

Therefore, the need for digital transformation has never been more pressing than it is today. In a world where data is the new gold, businesses that fail to adapt and evolve risk being left behind by their competitors. In order to remain competitive, companies must be able to leverage data and technology to streamline operations and improve the experience for users of any type.

Digital transformation is the key to achieving these goals, as it allows companies to first and foremost understand their current inefficiencies, automate manual processes, optimize their workflows, and gain real-time insights into their business operations. By embracing digital transformation, companies can gain a competitive edge and unlock new opportunities for growth and innovation.

However, and this is an important point to internalize, the need for digital transformation is not just about staying ahead of the competition. It's also about meeting the changing needs and expectations of users and customers in today's digital age. Both have come to expect seamless, personalized experiences across all digital channels, and companies that fail to deliver on these expectations risk losing their business to competitors who can consistently deliver to those demands. As such, digital transformation is not just an option for businesses, but is also imperative for survival in today's fast-paced and highly competitive market.

Rolling out a digital transformation effort

Rolling out digital transformation in a company that doesn't currently use it can seem like a daunting task, but with the right plan in place, it can be a smooth and successful process. And even if your company does currently use digital transformation practices, the steps will be the same, just – hopefully – understood by everyone and smoother to get moving and completed efficiently. It's just a matter of current perspective.

With that in mind, the first step is to assess the current state of the company's technology infrastructure, processes, and culture. This will help identify areas where digital transformation can have the most impact and where the greatest challenges may lie.

If you can make it happen, the assessment should be done by your core team of SMEs. Once complete, the next step is to develop a comprehensive digital transformation strategy that outlines the goals, objectives, and **key performance indicators** (**KPIs**) for the project. This strategy should be aligned with the company's overall business strategy and should consider the needs and expectations of stakeholders, employees, and customers.

The next step is to establish a governance structure and assign roles and responsibilities for the digital transformation project. This includes identifying a project team that will be responsible for implementing the strategy, as well as setting up processes for communication, decision-making, and risk management.

> **Note**
> A **Responsible, Accountable, Consulted, Informed** (**RACI**) matrix is a good tool to use for keeping track of all progress.

Most likely, a project manager or team will head this up, but, as we discussed in earlier chapters, if this is new to your company and you're introducing digital transformation, it may very well be up to your team to do this work too.

The next step is to identify the digital technologies and tools that will be required to support the digital transformation initiative. This may include software applications, hardware infrastructure, and cloud-based solutions. It's important to choose technologies that are scalable, secure, and easy to integrate with the existing technology stack.

After the required digital technologies and tools have been identified, a comprehensive assessment of the organization's current technological landscape is essential. This assessment involves evaluating the existing systems, infrastructure, and processes to determine their compatibility with the chosen technologies. It's crucial to identify any gaps or limitations that may hinder the smooth integration of new digital tools.

> **Note**
>
> Failing to conduct a comprehensive assessment of your organization's current technological landscape can lead to significant challenges and setbacks in the digital transformation journey. Without understanding the compatibility between existing systems, infrastructure, and processes, the integration of new digital technologies may encounter obstacles and setbacks that cause rework, which costs money and drains the budget.

Assuming everything is a go and you get the green light, the next step is to design and implement new processes and workflows that leverage the new digital tools, technologies, and processes. This may involve automating manual processes, redesigning customer experiences, and streamlining internal operations. Because we're now getting involved with workflows and internal operations, this is another opportunity for a change management team to be activated.

Finally, it's important to monitor and measure the success of the initiative using KPIs that align with the goals and objectives of the project. This will help identify areas for improvement and ensure that the project is delivering value to the company.

To give you a good place to start with KPIs to measure, here are some suggestions. While they may not all fit your particular needs, study these and use the ones that do. Each will require that you do the upfront work of determining exactly what to watch in each case:

- **Improved operational efficiency**: Measures the impact on the efficiency of your company's operations. It can be measured by tracking the time and resources required to complete key tasks before and after the implementation of digital tools and processes. You must determine the measurements and gather the metrics for them in the current work mode so that you have the baseline to compare to after the initiative is considered complete, or your first milestone has been hit.

- **Cost reduction**: Measures the impact on your company's bottom line. It can be measured by tracking the reduction in costs that were associated with manual processes being replaced, such as paper-based documentation and manual data entry. This can vary widely from company type to company type. A manufacturing company will have different criteria from a hospital, for example.

- **Increased revenue**: Measures the impact of digital transformation on your company's top line. Measure this by tracking the increase in revenue associated with new digital products and services, as well as improvements in customer experience and engagement. Notice here that we're suggesting both quantitative via the revenue, and qualitative via customer experience ratings.

- **Improved customer experience**: Measures the impact of digital transformation on your company's ability to deliver a seamless and personalized customer experience across all digital channels. It can be measured by tracking customer satisfaction scores, **Net Promoter Scores (NPSs)**, and customer retention rates.

- **Faster time to market**: Measures the impact on your company's ability to bring new products and services to market quickly. It can be measured by tracking the time it takes to develop, test, and launch new products and services before and after the implementation of digital tools and processes.

- **Improved employee productivity**: Measures the time and resources required to complete key tasks before and after the implementation of digital tools and processes, as well as employee satisfaction scores and retention rates. In initiatives that are concerned with improving internal metrics, this KPI is the strongest indicator of whether you're headed in the right direction. Indicators like this should be measured as the sprints release features so that fine-tuning adjustments can be made where needed.

Notice that more than a few of these KPIs can be combined. For instance, if you can prove that revenues increased while costs went down, that makes a much stronger case. Think about this list and come up with your own combinations to try or just to use in a design-thinking workshop so that your entire team and perhaps some stakeholders can be involved. With the right plan in place, digital transformation can be a powerful tool for driving growth, innovation, and competitive advantage.

Let's make this more evident with an example. I'm going to pull from my experience in the aviation industry and create a scenario for you, based on a real problem that my team and I solved.

Scenario: In facilities that repair jet engines, there are several roles. There's the production manager, who keeps track of all engines going through the plant and ensures that schedules are met. There are section chiefs, who report to the production manager from their respective engine module areas, such as gearbox. And – the role for this example – there's a kitting manager. This person's job is to ensure that all the parts are in the plant for every repair job that comes through.

Upon arrival at the plant, we begin an assessment to verify that all the work we did prior to getting onsite is still valid. Everything is okay, so we start interviewing each of the managers. Skip ahead to the kitting manager, and we find out that they have to use seven different software systems to do their job every morning. It takes them a full four hours – that's one half of their day – just piecing everything together from the disconnected systems. Because of this complexity, they sometimes miss important information but don't realize it until somewhere in hour three of the tedious work.

Finally, in hour five of their day, they are able to tell the production manager that only 60% of the work scheduled for tomorrow – they only work one day ahead if they have insight into what's about to hit the docks – will be able to happen. Rewind: there are days when something hits the dock that they didn't know about until that morning, and on occasions – rare but real – engines will arrive that they had no information about.

Because of this inefficiency of software systems, this plant gets fined every month for not delivering on time. The manager in charge of reassembling all the modules back into a full, working engine has to run around the plant looking for what they need to complete their work. If they are lucky, our kitting manager was able to get work completed in time to let them know by the middle of the day that they won't get everything that is expected.

So then, what would their day look like if they had an efficient system, one that was researched, designed, and built, and then worked into daily life through a well-thought-out and planned digital transformation initiative? Well, I'll tell you.

If the right digital transformation initiative were implemented in this plant, the kitting manager's day would be significantly improved, as would the efficiency of the entire facility. Rather than spending half of the day piecing together information from seven disconnected software systems, the kitting manager could use a single, streamlined software system that would automatically aggregate all the necessary data for them. This would free up several hours of their day, allowing them to focus on other important tasks, such as identifying potential bottlenecks and addressing them before they cause delays.

With a more efficient software system in place, the kitting manager would have real-time visibility into all incoming repair jobs, including information about the specific parts and tools needed for each job. They'd be able to track the status of each part, from the moment it arrives at the plant to the moment it's needed for a specific repair job. This would enable them to proactively address potential issues before they become major problems and would also help them to identify any missing parts or tools well in advance of when they're needed.

By using a more efficient software system, the kitting manager would be able to provide the production manager with accurate and timely information about the status of each repair job. They'd be able to give updates on progress, identify potential delays, and provide accurate estimates of when each job would be completed. This would help the production manager to plan and schedule work more effectively, resulting in fewer fine-incurring delays and less waste.

Overall, the implementation of a more efficient software system would result in significant cost savings for the facility, as well as improved customer satisfaction and increased revenue. By reducing the amount of time and effort required to manage the repair process, the facility would be able to deliver repairs more quickly and efficiently and would be able to meet or exceed customer expectations.

Our plan was to do just that, and after a few meetings with the plant director, whose job it was to ensure the safety of all his workers while meeting the demands of the customers who sent engines to them, we proposed a digital transformation plan that was accepted and rolled out over a six-month period.

We were able to reduce mistakes and material waste, improve the efficiency of the entire plant through five different module repair shops, improve their delivery rate according to contracts, which reduced their fines, and make our kitting manager so efficient that she was able to accept a promotion after training a new kitting manager. Which, of course, was very simple because the new system was very simple. Now that we've explored the importance of digital transformation and its impact on businesses, let's dive into how UX analytics can play a pivotal role in supporting your organization's digital transformation efforts.

How UX analytics supports the plan

UX analytics involves both contextual inquiry and the systematic collection and analysis of user data to gain valuable insights into user behavior, preferences, and interactions with digital systems. By leveraging UX analytics, you can empower your leaders and manager to make data-driven decisions to enhance user experiences, streamline processes, and drive successful digital transformations.

Implementing UX analytics tools enables organizations to gather both quantitative and qualitative data on user interactions with digital systems. This data can include metrics such as user engagement, conversion rates, task completion times, and user feedback. By examining these metrics, you gain a deeper understanding of how users interact with digital products or services, identify pain points, and uncover opportunities for improvement – that is, digital transformation initiatives.

Taking a user-centric approach, organizations can conduct user surveys, interviews, and usability testing to gain valuable insights into user needs, preferences, and pain points. This user feedback serves as a compass to guide the digital transformation journey. By understanding user perspectives, your team can prioritize enhancements and feature developments that directly address user challenges, resulting in more user-friendly and efficient digital experiences.

Through continuous monitoring and evaluation of user data, you can measure the effectiveness of the implemented changes and track the impact of the digital transformation initiatives. UX analytics enables you and your organization to identify trends, detect new pain points, and assess user satisfaction levels over time. This iterative approach allows for ongoing refinement of digital systems, ensuring they remain aligned with user needs and driving successful digital transformations.

By leveraging the power of UX analytics, organizations can make informed decisions, optimize user experiences, and achieve their digital transformation goals. The insights gained through UX analytics empower organizations to create digital products and services that resonate with users, drive engagement, and ultimately, lead to business growth and success.

Monitoring and evaluating user data

I think it's important to analyze this a bit more and explain why it's vitally important to you, your team, and ultimately, your company, as a digital transformation effort is widespread and will affect every user and role type in your organization. As we learned in *Chapter 11*, you can live by your data or you can die by your data.

Through continuous monitoring and evaluation of user data, organizations can measure the effectiveness of the implemented changes and track the impact of their digital transformation initiatives. UX analytics enables organizations to identify trends, detect new pain points, and assess user satisfaction levels over time, whether that's one week in evaluation of new features, or five years in evaluation of trends, both good and not so good. This iterative approach allows for the ongoing refinement of digital systems, ensuring they remain aligned with user needs and driving successful digital transformations.

By leveraging the power of UX analytics, you and your team can make informed decisions, optimize user experiences, and achieve your digital transformation goals. The insights gained through UX analytics empower organizations to create digital products and services that resonate with users, drive engagement, and ultimately, *lead to business growth and success*, which is usually the end goal. Every business wants to grow and continue being successful. And as one of my university professors put it, "*The number one goal of business is to stay in business.*" If you can't stay in business, there's no way to grow and achieve continued success.

This point bears repeating: an iterative approach empowers you to continuously refine your digital systems, recognizing and addressing the ever-evolving needs and expectations of your users. You know for a fact that in today's rapidly changing digital landscape, user expectations are in a constant state of flux, driven by advancements in technology, evolving market trends, and shifting customer preferences. The reality is that what satisfies users today may not meet their expectations tomorrow.

Therefore, it's crucial for organizations to embrace this dynamic nature of user needs and expectations. By adopting an iterative approach and actively incorporating user feedback into the refinement process, you can ensure that your digital systems remain agile and responsive to the evolving demands of your users. This ongoing commitment to adaptation and improvement is paramount in delivering exceptional user experiences and ensuring the long-term success of your digital transformation initiatives.

Statistics on user expectations

To solidify the earlier statement that "in today's rapidly changing digital landscape, user expectations are in a constant state of flux, driven by advancements in technology, evolving market trends, and shifting customer preferences," here are some statistics to give you a better understanding:

- 72% of customers say that they're willing to switch to a competitor if they have a bad customer experience

  - Source: Northridge Group. The study, which was conducted in 2019, surveyed over 1,000 consumers

- 52% of customers say that they expect a personalized experience from brands

  - Source: Salesforce. The study, which was conducted in 2018, surveyed over 10,000 consumers

- 42% of customers say that they're willing to pay more for a better customer experience

  - Source: PWC. The study, which was conducted in 2017, surveyed over 2,000 consumers

- 79% of customers say that they're more likely to do business with a company that offers a great customer experience

  - Source: Temkin Group. The study, which was conducted in 2018, surveyed over 2,000 consumers

- 90% of customers say that they're willing to recommend a company to others if they have a great customer experience

  - Source: American Express. The study, which was conducted in 2017, surveyed over 1,000 consumers

These statistics show that customer expectations are high and that businesses that can meet or exceed these expectations will be more successful. Businesses need to constantly innovate and adapt to the changing needs of their customers in order to stay ahead of the competition. Therefore, if you and your company can successfully meet the changing expectations of your customers, you'll be well positioned for success in the digital age.

Continuous monitoring for early problem detection

Through the systematic analysis of user data, organizations can uncover valuable insights that inform their decision-making process. By examining user behaviors, interactions, and feedback, organizations can identify patterns and trends that indicate the success or shortcomings of their digital transformation efforts. This data-driven approach provides a comprehensive understanding of how users engage with digital systems, highlighting areas of improvement and opportunities for innovation.

One of the key benefits of UX analytics is the ability to detect new pain points that may arise as a result of changes or updates to interfaces or backend parts of digital systems. By closely monitoring user interactions, organizations can identify areas where users encounter difficulties, frustrations, or inefficiencies. This early detection allows organizations to proactively address them, ensuring a smoother user experience and minimizing potential negative impacts on productivity, satisfaction, and overall success of digital transformation initiatives.

Also keep in mind that UX analytics enables organizations to assess user satisfaction levels over time, continually building a repository that allows for quick discovery when needed. By collecting user feedback, conducting surveys, and analyzing user sentiment, organizations can gauge how well their digital systems meet user expectations and requirements. This continuous assessment of user satisfaction provides valuable insights for making informed decisions about further refinements and optimizations to enhance the user experience and drive successful digital transformations. If you and your organization continue this over a longer period of time – I'm talking years here, 3 to 5 at a minimum – you now have a repository of information that can serve you in many ways, from marketing and sales efforts to new product rollouts and current state enhancements.

So then, in summary, the continuous monitoring and evaluation of user data through UX analytics empower organizations to measure the effectiveness of their digital transformation initiatives. By identifying trends, detecting pain points, and assessing user satisfaction levels, you and your company can make data-driven decisions to refine digital systems to ensure they remain aligned with user needs. This iterative approach fosters ongoing improvement, enabling organizations to deliver seamless, user-centric experiences and achieve successful digital transformations.

The leadership team

Proper leadership will ensure the success of a digital transformation effort. Effective leaders play a critical role in guiding and steering the organization through this transformative journey. They possess the vision, strategic mindset, and ability to rally teams toward a common goal. On the other hand, poor leadership can have detrimental effects, hindering progress and jeopardizing the entire initiative. The success of digital transformation relies heavily on the leadership team's commitment, support, and active participation.

If you're on the leadership team

If you're on the leadership team, bravo to you for embarking on this journey of discovery into UX analytics! Your decision to embrace digital transformation demonstrates your foresight and understanding of the ever-changing business landscape. Your teams need you now more than ever. They look to you to champion the cause of UX analytics-driven digital transformation because you possess the authority and influence to drive change and foster a culture of collaborative innovation.

By being the champion, you empower your teams to explore new possibilities and push the boundaries of what's achievable. You provide them with the resources, support, and encouragement necessary to adopt UX analytics practices and leverage data-driven insights. Your leadership sets the tone and creates an environment where teams feel inspired, empowered, and motivated to contribute their expertise and creativity toward the success of the digital transformation effort.

Your role as a leader is crucial in fostering a mindset of continuous improvement and adaptability. Digital transformation is not a one-time project; it's an ongoing journey of evolution and innovation. Your teams need you to foster a culture that embraces change, encourages experimentation, and values learning from both successes and failures. As a leader, you have the opportunity to shape the organizational mindset, enabling your teams to proactively adapt to the dynamic demands of the digital landscape.

Managing up

It's important to also understand that your role as a leader for digital transformation extends beyond your immediate team; it involves managing up through the organization and influencing upper management to understand the importance of a continued digital transformation effort. While this task may present challenges, there are strategies you can use to effectively communicate the need for ongoing digital transformation efforts and cultivate a culture that embraces change. Let's look at some of these:

- **Develop a compelling narrative**: Craft a clear and persuasive narrative that highlights the benefits of digital transformation for the organization as a whole. Tailor your message to resonate with upper management, focusing on the potential for increased efficiency, cost savings, competitive advantage, and improved customer experiences. Use data, case studies, and success stories to illustrate the positive impact that digital transformation can have on business outcomes.

- **Educate and build awareness**: Upper management may not be fully aware of the rapidly evolving digital landscape and the potential risks of falling behind. Invest time and effort in educating them about emerging technologies, market trends, and the changing expectations of customers. Organize workshops, seminars, and presentations to share insights, industry research, and success stories. Provide concrete examples of how digital transformation initiatives have driven growth and success for other organizations.

- **Align with strategic goals**: Demonstrate how digital transformation aligns with the organization's strategic goals and long-term vision. Show how it can drive innovation, enhance operational efficiency, and enable the organization to stay agile in a rapidly changing marketplace. Highlight the potential for new revenue streams, improved customer engagement, and enhanced competitiveness. Emphasize that digital transformation is not just a buzzword but a strategic imperative for sustainable growth and relevance.

- **Mitigate resistance**: Resistance to change is a common challenge in any transformation effort. Identify potential sources of resistance within upper management and address them proactively. Understand their concerns and hesitations and provide evidence-based explanations to alleviate their fears. Emphasize the importance of agility and adaptability in a digital-first world and reassure them that digital transformation is a continuous journey that requires ongoing effort and adjustment.

- **Foster a culture of innovation**: Encourage experimentation and provide opportunities for teams to showcase the value of digital transformation through pilot projects and proofs of concept. Celebrate successes and learn from failures, emphasizing the importance of a growth mindset and continuous improvement. And remember to always foster a safe environment where employees feel empowered to share ideas, take calculated risks, and challenge the status quo.

By managing effectively, you can help upper management understand the need for a continued digital transformation effort. It requires patience, persistence, and effective communication to convey the urgency and long-term benefits of embracing digital transformation. Remember that change takes time, and building a culture that embraces change is a gradual process. Your leadership and advocacy can pave the way for a successful and sustainable digital transformation journey throughout the organization.

Working with your teams

As a leader driving digital transformation, it's important to recognize that the success of your initiative relies on the support and collaboration of the teams you lead. In return, you can expect your teams to provide valuable support in various ways, contributing to the overall success of your initiatives.

Your teams bring a wealth of expertise and specialized skills to the table. They possess diverse skill sets, including UX design, development, data analysis, project management, strategy, and research skills, to name a few. Leverage their knowledge and empower them to apply their skills in driving the digital transformation forward. Encourage knowledge sharing and cross-functional collaboration, allowing team members to learn from one another and contribute their unique perspectives.

It's important to understand and recognize that innovation and creativity thrive within your teams. They understand the pain points, challenges, and opportunities within their respective areas of work. Lead and support an environment that values innovation and creativity, where team members feel empowered to propose and experiment with new ideas. Encourage them to think outside the box, challenge existing processes, and explore innovative solutions to improve user experiences and drive business outcomes.

They will also provide valuable feedback and insights based on their day-to-day interactions with digital systems, users, and data. Always encourage open communication channels that allow team members to share feedback, observations, and suggestions for improvement in a safe space, devoid of judgment and/or negative feedback. Seek their input during the planning and decision-making processes, as they can offer unique perspectives and identify areas for optimization. Emphasize the importance of data-driven decision-making and encourage teams to provide meaningful insights based on user analytics and UX research.

Adaptability and resilience are key qualities within your teams, and since digital transformation involves navigating through uncertainties, setbacks, and evolving requirements, you must do your best to foster a culture that embraces agility and resilience. You should embrace and support an environment where team members feel comfortable navigating ambiguity, adopting new technologies, and adapting to changing user needs. Encourage a growth mindset that embraces continuous learning, iteration, and improvement so that the best possible results come from the hard work that your team provides to support the organization.

Understand and champion this: collaboration is essential for success in digital transformation initiatives. Encourage your teams to work collaboratively, both within their own teams and across departments. Break down silos and facilitate cross-functional collaboration, where teams can share best practices, work on projects together, and learn from one another's experiences. Cultivate a sense of camaraderie and collective ownership of the digital transformation goals.

By setting the stage for a supportive environment and harnessing the potential of your teams, you can leverage their expertise, innovation, insights, adaptability, and collaborative efforts to drive successful digital transformation. Provide them with the necessary resources, recognition, and empowerment to excel in their roles and contribute to the overall vision. Together, as a cohesive and motivated team, you can navigate the complexities of digital transformation and achieve transformative outcomes.

Breaking down silos

A point made earlier talks about breaking down silos, and in some organizations, that's not the easiest thing to do. Old business values created them, and as we know, change is difficult for businesses as much as it is for individuals. Here's a short list of tactics for you to try if you face this type of challenge:

- Encourage frequent and transparent communication between teams and departments by creating and supporting open communication channels. Create platforms for sharing information, updates, and insights across the organization. Utilize collaborative tools and technologies to facilitate real-time communication and encourage cross-team collaboration.

- Break down silos by promoting cross-functional collaboration and teamwork. Encourage team members from different departments to work together on projects, initiatives, or problem-solving tasks. Facilitate cross-team meetings, workshops, and brainstorming sessions to foster a culture of collaboration and knowledge sharing.

- Align teams around shared goals and objectives that transcend departmental boundaries. Emphasize the importance of collaboration and cooperation in achieving these goals. Encourage teams to identify areas of overlap and interdependence and work together to find solutions that benefit the entire organization.

- Create an environment where team members feel safe to share ideas, opinions, and concerns without fear of judgment or reprisal. Foster a culture of trust by valuing diverse perspectives and encouraging open dialogue. Establish psychological safety as a core principle, where everyone feels comfortable taking risks, sharing feedback, and learning from mistakes.

- Break down silos by offering opportunities for team members to develop skills and gain knowledge outside their immediate areas of expertise. Encourage cross-training initiatives and job rotations that allow employees to gain a broader understanding of the organization and collaborate with colleagues from different departments. This cross-pollination of skills and knowledge enhances collaboration and breaks down silos.

By implementing these strategies, you, as a member of the leadership team, can effectively break down silos within the organization and spearhead a collaborative environment conducive to digital transformation. Breaking down silos enhances communication, collaboration, and knowledge sharing, enabling teams to work together toward shared goals and drive successful digital transformation initiatives.

As a leader in a company starting a digital transformation journey, you play a pivotal role in the success of the initiative and effort. Your commitment, vision, and support are instrumental in driving change, being a leading force for innovation, and ensuring the adoption of UX analytics practices. Embrace this opportunity to lead your teams toward a future where digital transformation becomes a catalyst for growth, resilience, and enhanced user experiences. Your teams are counting on you to navigate this transformative journey, and with your guidance, they can realize the full potential of digital transformation.

Working with your peers

To effectively evangelize digital transformation and emphasize the importance of cross-functional teamwork to peers, you can employ a strategic and collaborative approach.

Here are two suggestions on how to approach your peers to help them understand so that they can lend support from their place and perspective within the organization.

Firstly, focus on building awareness and understanding by highlighting how UX analytics can provide the foundation for digital transformation and its impact on the organization's overall success. You can organize workshops, presentations, or informal discussions to share success stories and case studies that demonstrate the positive outcomes achieved through digital transformation initiatives. By showcasing real-world examples, you'll help your peers grasp the tangible benefits, such as increased efficiency, improved customer experiences, and enhanced competitiveness. It's important to emphasize that digital transformation is not merely a technological shift but a strategic imperative that requires cross-functional collaboration.

Secondly, you should engage your peers in meaningful conversations and demonstrate the interdependencies between different functions and departments. Facilitate discussions and workshops that bring together representatives from various teams to identify shared challenges and opportunities. By encouraging open dialogue and active participation, you can guide your peers toward a collective understanding of the need for cross-functional teamwork in achieving successful digital transformation. It's crucial to emphasize that digital transformation initiatives are not isolated efforts but require a collaborative approach that leverages the diverse skills, expertise, and perspectives of different teams. By promoting a shared vision and creating a sense of ownership, you can inspire your peers to actively support and participate in cross-functional initiatives, creating and nurturing a culture of collaboration and driving the organization's digital transformation journey.

Supporting your leadership team

So, let's turn this around now and consider what it takes to be a good teammate in supporting your leadership team. This will sometimes involve managing up, just as the leaders do with upper management, and it will also incorporate your ability to listen, take direction, and work toward the goals set by your leaders.

Let's start with three simple tactics to get you warmed up and in the right frame of mind to internalize the importance of being that teammate that every leader wants to guide and coach:

- Always be proactive and take the initiative to familiarize yourself with the vision and objectives set. Understand how your work in UX analytics contributes to the broader digital transformation efforts. Align your priorities and projects with the strategic direction set by the leaders to ensure your work is in sync with their expectations.

- Regularly communicate your findings, insights, and recommendations to the leadership team in a clear and concise manner. Present data-driven insights and actionable recommendations that demonstrate the impact of UX analytics on the organization's goals. Provide updates on key metrics, user feedback, and trends that can help inform decision-making and shape the digital transformation strategy.

- Actively collaborate with other teams and departments to foster cross-functional partnerships. Seek opportunities to work together, share knowledge, and contribute to joint initiatives. By collaborating with colleagues in areas such as design, development, and marketing, you can help bridge gaps and ensure a holistic approach to digital transformation. Offer your expertise in UX analytics and contribute to the collective effort of achieving successful outcomes.

Okay? Good start, right? Now, let's dig into it.

As a UX analytics pro in your organization and on your team, your role in helping leadership craft the strategy of the digital transformation effort is invaluable. You have a unique perspective and expertise that can greatly influence the direction and success of the transformation. Let me explain why it's crucial for you to actively contribute to strategy development.

First and foremost, your deep understanding of user behavior, needs, and preferences gives you a user-centric perspective that's crucial to project success. By actively participating in strategy discussions, you can ensure that the digital transformation efforts are aligned with user needs and focused on delivering exceptional user experiences. Your insights and input will help shape a strategy that truly puts the end users at the center, increasing the chances of success.

In addition, your proficiency in analyzing and interpreting user data enables you to provide data-driven recommendations to leadership. This data-driven decision-making approach ensures that you're developing a strategy that's grounded in real user behavior and feedback. Your expertise will help leadership make informed decisions and prioritize initiatives that have the highest potential for impact while being cost-effective to the company. Your contributions will be instrumental in maximizing the outcomes of the effort.

And understand – I mean *really* get this – that your involvement in strategy development helps identify potential risks and challenges early on. Your analysis of user data and UX research uncovers areas of improvement, anticipates user needs, and identifies usability issues. By sharing these insights with leadership, you'll help them make strategic decisions that mitigate risks and increase the likelihood of success. Your input will be critical in guiding resource allocation, ensuring that investments are focused on initiatives that have the greatest potential for achieving desired outcomes.

Lastly, your active participation in strategy discussions fosters collaboration and alignment across different teams and departments. You bridge the gap between user insights and strategic decision-making, creating a cohesive and holistic approach to digital transformation. By actively engaging in these discussions, you facilitate a shared vision that focuses efforts to guarantee that everyone is working toward the same goals. Your contribution strengthens the overall effectiveness of the digital transformation effort and increases the likelihood of achieving the desired outcomes.

Remember, as a UX analytics team member, you have a unique role to play in shaping the strategy of the digital transformation effort. Your user-centric perspective, data-driven insights, and collaboration skills are essential. By actively participating and sharing your expertise, you'll make a significant impact on the success of the transformation. Your contribution matters, and your role as a trusted advisor to leadership is crucial in driving the organization toward a successful digital transformation. In essence, be the leader your company needs you to be.

Being a supportive teammate

We all have peers, and we all support those peers in strong teams. Just like you need a solid understanding of how to support your leadership team, you also need a solid understanding and path forward in working with and supporting the understanding and efforts of your teammates.

I like to look at this through the lens of the theory of constraints but at the level of the team. *Figure 13.1* shows a simple chart to illustrate a three-person team consisting of a researcher, a designer, and a developer, and the interdependencies associated with this team. And a quick reminder that the theory of constraints analyzes connected systems for potential bottlenecks.

Role	Directly Influences	Indirectly Influences
Researcher	Designer	Developer
Designer	Researcher, Developer	
Developer	Designer	Researcher

Figure 13.1 – Influence diagram

And here's what this looks like as a mind map, or, if you like, an entity relationship diagram:

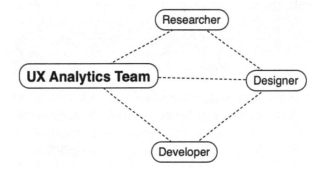

Figure 13.2 – Mind map of the influence relationship

Each role – researcher, designer, and developer – belongs to the team, so they all support the team and the overall efforts. Note that the designer is directly tied to both the researcher and the developer. This is because the designer must interpret the work of the researcher when doing solution discovery and must also relay the best information to the developer. Let's dive deeper into this for a firm understanding of how these relationships work.

Unlocking the relationships between the roles

The researcher's role is to gather insights, conduct user studies, and analyze data to understand user behavior, needs, and preferences. Their findings provide a solid foundation of knowledge that informs the design and development process. The researcher's work is closely tied to the designer as they provide valuable insights and data that shape the design decisions. The designer relies on the researcher's work to gain a deep understanding of the user's perspective, enabling them to create user-centric solutions. The researcher's insights help the designer identify pain points, uncover opportunities, and make informed design choices.

> **Who and what is the researcher dependent on?**
>
> The users and the data gathered. They must get to the right people and analytics data gathered in order to portray current state situations to pass along.

The designer, in turn, takes the insights provided by the researcher and translates them into intended, workable solutions. They leverage their creative skills, knowledge of user experience principles, and understanding of design best practices to craft user interfaces and experiences that align with user needs. In essence, the designer is the bridge between the researcher and the developer, as they interpret the researcher's findings and translate them into actionable design specifications. They ensure that the developer has a clear understanding of the design requirements, effectively communicating the best information to bring the solution to life.

Finally, the developer takes the design specifications provided by the designer and turns them into functional digital products or applications. They're responsible for coding, programming, and implementing the design. The developer relies on the designer's expertise to understand the intended user experience and effectively translate it into a working product. Through close collaboration, the developer seeks clarification from the designer when needed, ensuring that the implementation aligns with the desired design outcomes and user expectations.

In this symbiotic relationship, the researcher, designer, and developer work together as a cohesive team. The researcher provides valuable insights and data, shaping the design decisions. The designer translates those insights into well-crafted designs, effectively communicating with both the researcher and the developer. The developer brings the design to life, relying on the designer's guidance and expertise. Together, they create user-centered digital solutions that drive digital transformation efforts forward. This collaboration and synergy among the team members ensure that the user's needs and expectations are at the core of the transformation process.

Potential bottleneck #1 – the researcher

When/if the researcher gets behind or fails to provide conclusive results from their work, it can lead to significant bottlenecks in the solution discovery process for the designer. Let's explore some of the potential bottlenecks that may arise in such situations:

- **Lack of user insights**: The designer heavily relies on the researcher's findings to gain a deep understanding of user behavior, needs, and preferences. If the researcher falls behind or fails to provide clear and comprehensive results, it creates a gap in the designer's knowledge base. Without sufficient user insights, the designer may struggle to make informed design decisions and may resort to guesswork or assumptions, compromising the effectiveness of the solution discovery process.

- **Inadequate problem identification**: The researcher plays a crucial role in identifying pain points, uncovering user needs, and highlighting areas that require improvement. When the researcher doesn't effectively deliver the results of their research, it becomes challenging for the designer to accurately identify and define the problem that needs to be solved. This lack of clarity can lead to a vague design direction and hinder the effectiveness of the solution discovery phase.

- **Reduced alignment with user expectations**: The purpose of research is to understand user expectations and relay these in a fashion that allows the designer to align the design solutions accordingly. If the researcher fails to prove the results of their research or falls behind schedule, it delays the designer's ability to create user-centered solutions. This delay may result in a disconnect between the final design and user expectations, leading to subpar user experiences and reduced effectiveness of the digital transformation efforts.

- **Impaired collaboration and communication**: Research findings serve as a common language between the researcher and the designer. The properly articulated findings facilitate effective collaboration and communication, ensuring that the designer's solutions are grounded in user insights. However, when the researcher lags behind or doesn't provide conclusive results, it hampers the collaboration between the researcher and the designer. Miscommunication, unclear requirements, and a lack of shared understanding may arise, impeding the solution discovery process and slowing down progress.

To mitigate these bottlenecks, it's essential for the researcher to stay on track, deliver research results in a timely manner, and provide clear and actionable insights to the designer. Open and regular communication between the researcher and the designer is crucial to address any challenges or delays proactively. By maintaining a seamless flow of information and ensuring the availability of user insights, the researcher can support the designer in effectively conducting solution discovery, leading to more successful digital transformation outcomes.

Potential bottleneck #2 – the designer

So then, what if the designer misinterprets information from the researcher, or maybe just simply relays something to the developer that they feel will make things easier?

When the designer misinterprets information from the researcher or relays inaccurate or misleading details to the developer, it can lead to significant challenges and potential setbacks in the digital transformation process. Let's explore some of the implications that may arise from such situations:

- **Misaligned design solutions**: If the designer fails to fully understand or properly interpret the insights provided by the researcher, it may lead to a mismatch between the intended design outcomes and the actual user requirements. As a result, the digital solution may not effectively solve the identified problems, leading to suboptimal user experiences and decreased user satisfaction.

- **Inefficient development process**: If the designer relays inaccurate or misleading information to the developer, it can significantly impact the development process. The developer may receive incomplete or incorrect design specifications, leading to confusion, rework, and delays. This miscommunication can disrupt the development workflow, leading to increased time and effort required to rectify the issues, potentially impacting project timelines and, ultimately, the budget set for the project.

- **Compromised user experience**: If the designer conveys simplified or incorrect details to the developer in an attempt to make things easier, it can result in a product that doesn't meet the users' expectations or fails to fulfill their needs. This can negatively impact user satisfaction, hinder adoption rates, and undermine the success of the digital transformation initiative.

- **Strained collaboration and communication**: Misinterpretations or inaccuracies in information exchange between the designer and developer can strain collaboration and communication within the team. Effective collaboration relies on clear and accurate information sharing, ensuring that everyone is aligned in their understanding of user needs and design goals. Misinterpretations or misleading information can hinder this collaboration, leading to inefficiencies and decreased team cohesion.

To reduce or remove the likelihood of these challenges, it's crucial for the designer to maintain open lines of communication with both the researcher and the developer. Regular check-ins, clarifications, and seeking feedback can help ensure that the information is accurately interpreted and relayed. Collaboration and iterative feedback loops between the designer, researcher, and developer are essential to address any misunderstandings and ensure that the design solutions align with the user insights. Clear communication channels and a shared understanding of the project goals can help prevent misinterpretations and support a more effective digital transformation process.

Potential bottleneck #3 – the developer

During the development phase of a digital transformation effort, several bottlenecks can potentially hinder progress and impact the successful execution of the project. Here are some common bottlenecks in development to be aware of:

- **Inadequate communication and collaboration**: When there's a lack of clear and timely communication, misunderstandings can arise, requirements may be misinterpreted, and development tasks can become delayed or misaligned. It's essential to ensure effective communication between design and development, encourage regular updates and feedback, and promote collaboration to overcome this bottleneck.

- **Resource constraints**: Limited availability of resources, such as skilled developers, technology infrastructure, or budget, can pose significant bottlenecks in development. This can lead to delays, compromised quality, or an inability to tackle complex development challenges effectively. It's crucial to assess resource requirements early in the project and ensure that the necessary resources are allocated appropriately to avoid bottlenecks.

- **Technical complexities and dependencies**: Development efforts often involve complex systems, integration challenges, or dependencies on external tools or services. Issues such as compatibility conflicts, unresolved dependencies, or architectural constraints can impede progress and require additional time and effort to overcome. Conducting thorough technical planning, addressing potential challenges proactively, and seeking appropriate expertise can help mitigate these types of delays.

- **Scope creep and changing requirements**: Uncontrolled scope creep and constantly changing requirements can disrupt development efforts. When project scope and requirements evolve without proper management, they can lead to delays, increased development efforts, and a lack of clarity regarding project goals. Implementing effective change management processes, regularly reassessing requirements, and ensuring stakeholder alignment can help mitigate the impact of scope creep and changing requirements.

- **Lack of automated testing and quality assurance**: Inadequate or inefficient testing and quality assurance practices can result in bottlenecks during development. Manual testing processes, limited test coverage, or a lack of automated testing frameworks can lead to inefficient testing efforts, longer feedback cycles, and higher chances of introducing bugs or issues. Investing in robust testing strategies, implementing automated testing tools, and promoting a culture of quality assurance can help alleviate this bottleneck and the problems it can cause.

By identifying and addressing these potential bottlenecks, development teams can proactively navigate challenges, improve efficiency, and maintain momentum throughout the digital transformation journey.

Tackling the need for new/additional changes

When new research reveals insights or findings that weren't previously identified during the initial research phase, it becomes crucial to effectively communicate and push these changes up to the development team immediately. Here are some best practices for introducing research-driven changes during the development phase:

- **Document and consolidate the new findings**: Compile the new research findings in a clear and concise manner. Document the key insights, supporting data, and any implications or recommendations resulting from the research. Present the information to your designer exactly as you did in the first iteration. Always establish and follow presentation rules and best practices to maintain consistency.

- **Share the research findings with the relevant stakeholders**: This should include project managers, as well as the designers, developers, and other team members involved in the development process. Share the research findings through meetings, presentations, or well-written reports to ensure that everyone is aware of the new insights.

- **Highlight the impact on the project**: Clearly articulate the implications of the new research on the project and explain how it affects the current development work. Emphasize the value and significance of the new findings in terms of improving user experience, addressing pain points, or enhancing the overall success of the digital transformation effort. Connect the dots between the research and the potential benefits of the project.

- **Evaluate the impact on project timelines and priorities**: Assess the impact of the research-driven changes on project timelines, priorities, and resource allocation. Work closely with the design and development teams to determine the feasibility of incorporating the changes without significantly disrupting the existing development process. Collaboratively discuss and negotiate any necessary adjustments to project plans, ensuring that the revised approach is realistic and aligned with the project goals.

- **Provide support and resources**: Offer support to the design and development teams in implementing the research-driven changes. This may include suggesting additional resources or facilitating work toward any necessary adjustments to design specifications or requirements. Ensure that the necessary tools, documentation, and guidance are provided to aid the successful integration of the research findings into the development process.

Occurrences like this can happen, so it's smart to be aware and know what to do if and when they arise. Sometimes, schedules are so tight that management will force the beginning of development before all the proper research protocols have been followed and the problem discovery reports have been read out and turned over to the design team.

And sometimes it's just good business to get something started in the way of light prototypes so that initial user testing can begin. There are many contingencies involved when doing this, so be sure that all teams are involved and that everyone – especially the stakeholder – is on board with proceeding like this. If everyone is aware of the possible pitfalls, then no one is surprised when further research proves another course of action is warranted, or even contradicts current thinking.

By following these practices and the steps introduced, you can effectively communicate and push research-driven changes up to the development team, while fostering a collaborative environment that embraces new insights and promotes continuous improvement throughout the digital transformation journey.

Final notes on these dependencies

As you can see, collaboration and synergy among the researcher, designer, and developer are essential for the successful delivery of quality software in a digital transformation effort. By adopting the following best practices, these team members can work together effectively.

First, (and as you can see, this is a strong trend in this chapter) establish a culture of open and continuous communication. Encourage regular updates and feedback sharing to ensure a shared understanding of project goals, requirements, and user needs. Transparent communication fosters collaboration, prevents misunderstandings, and addresses potential issues early on. While the researcher and developer will only occasionally work directly together, it's imperative that they know how to communicate and understand each other's language.

Second, promote shared knowledge and understanding. Emphasize the importance of cross-functional learning between the researcher, designer, and developer. The designer should have a clear understanding of the research findings, and the developer should be familiar with the design objectives. This shared knowledge aligns everyone toward a common vision, leading to better decision-making and improved outcomes.

Third, embrace an iterative and feedback-driven approach. Make iterations and feedback loops a fundamental part of the collaboration process. Regularly review and refine designs, incorporate user feedback, and adjust development efforts based on research findings. This iterative process uncovers valuable insights, validates assumptions, and ensures the final product meets user expectations effectively. If you work in Agile, this will be familiar to you, and if you work in a waterfall environment, it's a practice you'll need to implement to the best of your ability.

Additionally, encourage collaborative problem-solving when challenges arise. Bring together the researcher, designer, and developer to collectively brainstorm solutions. Leveraging diverse perspectives and expertise will always lead to innovative problem-solving approaches. Use design thinking with an impartial facilitator where possible to lead the sessions.

Always – always – maintain a strong focus on empathy and user-centricity throughout the collaboration process. The researcher's insights into user behavior and needs should inform design decisions, while the developer strives to create a seamless and intuitive user experience from the solution discovery efforts of the designer. By keeping the end user at the center of the collaboration, the team can create digital solutions that truly meet their needs and deliver value.

And finally, foster a culture of respect and appreciation for each team member's role. Recognize and value the unique contributions of the researcher, designer, and developer. Encourage open dialogue, provide opportunities for skill development, and celebrate achievements to create a positive and supportive working environment.

Summary

In this lengthy and information-packed chapter, we got deeply involved in the world of digital transformation, exploring its significance and key components. We began by highlighting the need for digital transformation in today's fast-paced world, where data is the new gold and companies must adapt to survive. We emphasized the importance of leadership involvement in driving successful digital transformation efforts, advocating for a culture that embraces change and values continuous learning.

UX analytics has emerged as a crucial tool for guiding digital transformation initiatives. We discussed its role in measuring the effectiveness of implemented changes, tracking the impact of transformation efforts, and identifying trends and user pain points. The iterative nature of the discipline allows for ongoing refinement of digital systems to meet evolving user needs and drive successful transformations.

We also explored the interdependencies between the researcher, designer, and developer within the digital transformation journey. By highlighting the importance of effective collaboration and communication, we underscored how these roles support each other and contribute to the overall success of the project. We discussed potential bottlenecks and challenges that may arise at each stage and provided guidance on overcoming them.

And last, but certainly not least, we emphasized the significance of teamwork and support from both leaders and team members. Leaders were encouraged to champion the cause of digital transformation, foster a culture of change, and actively engage with peers at their leadership level to drive understanding and support. Team members were urged to contribute their expertise, support leadership efforts, and actively participate in shaping the strategy and execution of the transformation.

By combining leadership guidance, effective collaboration, and a user-centric approach, organizations can navigate the complexities of digital transformation and unlock new opportunities for growth and success in the digital age.

In the upcoming chapter, we'll get into content strategy and analysis, learning the differences between content and data, how to use taxonomy to organize and understand content structure, and to wrap it up, we'll bring everything full circle to how UX analytics drives content context and creation.

14
Content Strategy and Analysis – an Analytics-Based, User-Centric Approach

Just as data and research form the foundation of UX analytics, content serves as the visible structure that users interact with, making it the face of your company. Effective content strategy and analysis, supported by UX analytics, are vital for delivering a seamless user experience, driving engagement, and achieving business objectives.

Content forms the visible structure that represents your company to users, and its effective management, supported by UX analytics, is crucial for digital business success. By harnessing the power of content strategy and analysis, you can optimize the user experience, drive engagement, and align your content with the needs and expectations of your target audience.

You must also – always – consider storage and access to content. These are critical aspects of digital business operations. Efficient content management systems, databases, and repositories enable the secure storage, retrieval, and delivery of content to end users. Outside the scope of this book, it's an important topic and should be studied with the right team of people so as to ensure easy and secure access and management.

In this chapter, we're going to cover the following main topics:

- Understanding the difference between content and data
- Using a taxonomy to understand content structure
- Utilizing your analytics work for content strategy

Understanding the difference between content and data

Content encompasses a wide range of elements, including text, images, videos, and interactive elements. It serves as the medium through which information, messages, and experiences are conveyed to users. Capturing and organizing content in a structured manner ensures its integrity, consistency, and relevance across various touchpoints and channels. With the help of UX analytics, you can gain valuable insights into how users interact with different types of content, enabling you to make data-driven decisions that optimize the user experience.

Data, on the other hand, serves as one of the fundamental building blocks of content. It provides the raw material and insights that shape the creation, curation, and delivery of meaningful content experiences. It comes from your UX analytics work and represents your users' interactions with your application – whether external customer-facing applications, or internal applications used by your company's employees.

Data fuels the understanding of user behaviors, preferences, and needs, enabling you to make informed decisions when crafting content strategies. Through UX analytics, you can leverage data to uncover valuable patterns, trends, and correlations that inform content creation, optimization, and personalization efforts. And by harnessing the power of data-driven insights, you ensure that your content resonates with your target audience, drives engagement, and delivers value at every touchpoint.

Content performance analysis

UX analytics can also provide insights into content performance, allowing you to assess the impact of your content strategy on user engagement, conversion rates, and overall business goals. By understanding how users consume and interact with content, you can refine your content strategy, tailor experiences to different user segments, and continually improve the relevance and effectiveness of your content.

You and your team can also identify potential bottlenecks that impede user engagement. For instance, if you notice high bounce rates on a specific page, UX analytics can help pinpoint the specific pain points that cause users to abandon the site there. Armed with this information, you can implement targeted improvements, swap out the content for something that's perhaps more relevant, and remove any barriers that hinder seamless navigation.

Continuous improvement is a key factor in any successful content strategy, and UX analytics plays a vital role in this iterative process. By regularly monitoring user behavior and analyzing UX data, you can track the performance of your content over time and identify areas for improvement. This could involve tweaking headlines to increase click-through rates, experimenting with different layouts or visuals, or adjusting the language and tone to better resonate with your target audience.

Adjusting language and tone

Even if you're years into an application or website, keeping the correct tone can be a challenge. If your company pivots or uncovers a new market segment they'd like to enter, you can be in new territory that's unfamiliar and therefore need to study more. This can be true even for new products you're introducing to existing customers. And let's not forget that teams change, people leave, and synergies get altered with new managers and directors.

In such cases, you can find yourself in need of adjusting the language and tone to effectively communicate with your new and refined target audience. To achieve this, it is essential to have a deep understanding of who your target audience is and what language and tone resonate with them. Here are some key considerations for aligning your language and tone with your audience:

- **Define your target audience**: Begin by clearly defining your target audience based on demographics, psychographics, and other relevant factors. Consider their age, gender, education level, interests, values, and preferences. This information will provide insights into their communication style and help you tailor your language and tone accordingly:

  - Segmentation can be incorporated into the process of defining the target audience by further categorizing the audience based on specific characteristics or attributes. While demographics, psychographics, and other relevant factors provide a general understanding of the audience, segmentation allows for a more detailed and focused approach.

  - Segmentation involves dividing the target audience into distinct groups or segments based on shared characteristics such as age, gender, education level, interests, values, and preferences. By segmenting the audience, you can create more targeted and personalized communication strategies.

- **Conduct audience research**: Conduct thorough audience research to gain a deeper understanding of their language preferences and communication styles. This can involve surveys, interviews, focus groups, or analyzing social media conversations related to your industry or niche. Pay attention to the words, phrases, and tone they use in their conversations to guide your content creation.

- **Use appropriate language**: Once you have a clear understanding of your target audience, choose language that aligns with their preferences. This may involve using formal or informal language, technical or simple terms, and industry jargon or plain language, depending on the characteristics of your audience. The language you use should feel natural and relatable to your target audience, making it easier for them to connect with and understand your content.

- **Consider cultural nuances**: If your target audience consists of people from diverse cultural backgrounds, be mindful of cultural nuances in language and tone. Avoid using idioms, slang, or technical jargon that may not be universally understood or could potentially offend certain cultural groups. Research cultural norms and sensitivities to ensure your content is inclusive and respectful.

- **Reflect your brand identity**: While it's important to match your language and tone to your target audience, it's also crucial to maintain consistency with your brand identity. Your brand has its own personality and values that should be reflected in your content. Strive for a balance between aligning with your audience and staying true to your brand voice.

- **Test and iterate**: Regularly monitor and analyze the performance of your content to see how well it resonates with your target audience. Pay attention to engagement metrics, feedback, and conversions to gauge the effectiveness of your language and tone. Use this feedback to make iterative improvements and refine your content strategy over time.

To create impactful content, it's vital to understand and connect with your target audience. By aligning your language and tone with their preferences, you can make them feel understood, engaged, and motivated to take action, and this establishes a strong connection that builds lasting relationships. Armed with these insights, you can continually refine your content strategy, make informed decisions, and enhance the relevance and effectiveness of your content over time.

Key UX analytics metrics for content

Page views, time on page, bounce rate, and conversion rates are key indicators of the success or failure of your content performance, and therefore your content strategy. While page views are an indicator of how well you've designed your taxonomy and made topic, section, and link titles relevant to users, time on page and bounce rates are two of the strongest indicators of content strategy.

Page views

If a user can't find what they want or need, they can't visit the pages that contain the information. And that's considering that you actually have it available. As a quick example, let's say you're a medical information company and someone comes in searching for information on cholesterol. If you have 30 pages of information and 50% of your visitors only access three of the pages, those aren't good statistics. Three pages is only 10% of your content on the subject and if only 50% of your visitors see the pages, you're down to 5% page views on your cholesterol content.

So, back to that taxonomy design mentioned earlier: if people can't find it, they don't get the benefit of your company's expertise in that subject and field. To take this a bit deeper, let's consider that by monitoring page views, you can assess which promotional channels and tactics are generating the most views. This information enables you to refine your promotion strategy, focus on channels that deliver higher page views, and allocate resources accordingly.

By tracking page views over time, you can identify content that experiences a decline in views or fails to attract significant traffic. This indicates the need for content optimization, such as updating outdated information, improving search visibility, or enhancing the user experience. Optimization efforts can help revitalize underperforming content and improve its visibility to attract more page views.

Time on page

Time on page is a critical metric that offers valuable insights into user engagement with your content. When users spend a substantial amount of time on a page, it indicates that your content is successfully capturing their attention, resonating with their interests, and delivering value. A longer time on page suggests that users are actively consuming the information, diving deeper into the content, and potentially finding it helpful or engaging.

This metric can help you identify which content pieces are performing well in terms of holding user interest and providing meaningful experiences. By understanding the factors that contribute to a longer time on page, you can optimize your content strategy to create more engaging and valuable content that keeps users hooked and encourages further exploration.

Bounce rate

On the other hand, high bounce rates can be a cause for concern as they signify potential issues with content relevance, quality, or user experience. When users leave a page without further interaction, it suggests that the content may not have met their expectations or addressed their needs. It could indicate that the content wasn't relevant to their search query or that it failed to provide the information they were seeking.

High bounce rates can also point to issues with the quality of the content, such as poor readability, lack of visual appeal, or outdated information. Additionally, user experience factors such as slow page load times, intrusive popups, or difficult navigation can contribute to high bounce rates. By analyzing and addressing high bounce rates, you can identify areas for improvement, refine your content strategy, and enhance the overall user experience to better engage and retain your audience.

> **Idle versus active bounce**
>
> An idle bounce is considered to be any block of time over 30 minutes where there's no interaction from the user. An active bounce is caused by any action that takes the user away from the current page.

Conversion rate

Conversion rates are a critical metric for evaluating the success of your content strategy. This metric measures the percentage of users who complete a desired action, such as subscribing to a newsletter, filling out a form, or making a purchase.

According to *ranktracker.com*,

> *"The average conversion rate for websites is approximately 2.35% across all industries. Though certain industries will have higher conversion rates, others will have lower. A conversion rate of 11% is recorded on the best-performing websites."*

A high conversion rate signifies that your content is effectively persuading and motivating users to take the desired action, demonstrating the effectiveness of your content strategy. Conversely, low conversion rates may indicate areas that require optimization, such as call-to-action placement, messaging, or overall user experience.

Keep in mind, however, that only a certain percentage of your application or site will be content that indeed seeks a desired action. If you have information-only content, there's no action required, so you'd be looking at time on page or bounce rates to gauge whether that content is working or not.

Using a taxonomy to understand content structure

One of the first questions to be asked when tackling any new analytics project is "Where's your taxonomy?" This question catches a lot of people off guard when asked because a lot of companies don't consider the benefits that having one will offer. This will be especially true if you work for a consulting firm, or are on a professional services team that works with many customers per year.

In my time spent with a large software company in New York City, my dig into the website taxonomy brought up thousands of outdated pages, broken links, dead ends, and hidden content that very few people even knew existed. I captured as much as I could find and transferred it all to a taxonomy tool, printed it, and put it on the wall in a large conference room so that everyone could see what a total mess it was.

Out of the 20-30 people who reviewed it, only 2 were aware of what a true disaster it was, yet they were powerless to do anything about it because of constraints – time, backend, architecture, budget, you name it – hence it hadn't been tackled in 15 years.

I tell you this to illustrate what you may one day face, if you aren't currently. And to point out that if you don't have a taxonomy, make it now. It will help solve content problems for you and your customers/ users because whether internal or external, content is content and it must be managed properly. Your UX analytics work can solve a lot of problems.

Here's a simple taxonomy example to whet your appetite:

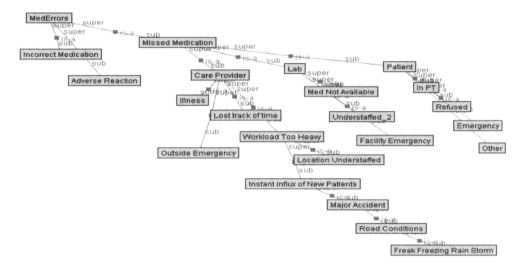

Figure 14.1 – A simple taxonomy example

The starter category, as you can see, is **MedErrors**. This is based on work my team and I did in the medical field a few years ago and serves to shed light on a few important aspects of taxonomies. **Missed Medication** is the deepest category, and **Incorrect Medication** is the shallowest. Each of these labels describes reasons why the *MedError* occurred and each is now a content type.

Let's walk down a leg:

A MedError occurred because of a Missed Medication. It's a lab error because of the following:

- The med was not available:

  - Because the pharmacy was understaffed:

    - Because of a facility emergency

So, a facility emergency kicked off a chain reaction of problems that ultimately caused a med error. Study the other legs and categories and then we'll skip to a content taxonomy example. I started you off with this one because it's qualitative data and helps you grasp this concept quicker – if you don't already practice it. If you do, then yeah, you're with me all the way and most likely evangelizing the power of taxonomies wherever you go. We're kin.

Before we skip to content, let's look at the major aspects of taxonomy usage for your analytics practice:

- **Terminology**: A taxonomy is a system of terms and their relationships. The terms in a taxonomy can be anything from words to phrases to concepts.

- **Hierarchy**: Taxonomies are hierarchical, meaning that the terms are arranged in a tree-like structure, or if you prefer, a parent/child relationship. This allows users to navigate the taxonomy and find the information they are looking for quickly and easily.

- **Labels**: Taxonomies often use labels to identify the terms. Labels can be human-readable or machine-readable.

- **Properties**: Taxonomies can also use properties to further describe the terms. Properties can be used to store additional information about the terms, such as their definitions, synonyms, or examples.

- **Mappings**: Taxonomies can be mapped to other taxonomies or to external data sources. This allows users to find related information across different systems.

In addition to these technical aspects, the benefits of using taxonomies for content strategy are numerous, and actually quite obvious once you understand the concept. Taxonomies can help you to do the following:

- Organize your content:

  - What types of content go in the Help category?

- Improve the discoverability of your content:

  - What top-level terms do your customers understand and need to access most frequently?

- Make your content more accessible to users:

  - What have you done to ensure that your users understand your terms, categories, and wayfinding paths?

- Increase the reusability of your content:

  - What content do you have for product *xyz* that can help you with a good starting point for *xyz2*?

- Improve the quality of your content:

  - What words are misspelled? What information is now outdated or irrelevant to this market?

- Reduce the cost of managing your content:

  - All the preceding points lead to this. When it's easy for your users to find and use your content, and it's easier for you to create and insert new content, the cost is reduced, and everyone's productivity increases.

Here's another real-world example of poor, well, really, *no* taxonomy or terminology thought or practice. I've seen this in several companies so I'm not calling out anyone in particular here, but consider the following:

- Widget – from marketing
- WIDGET – from engineering
- widget – from product
- wiget – from someone in accounting

While everyone thinks they're talking about the same thing, the software and backend systems treat all the terms in the preceding list as different entities. In a database, they are four different pieces of metadata. So, when a spreadsheet comes in with 1,000 "Widgets" and 600 "WIDGETs," they are not represented as 1600 of the same item, and that's a huge problem for inventory, accounting, sales… Right? While it's probable that they all represent the exact same physical item, according to the data, they're all different. I've seen this problem more times than I can count.

Now, how about completely different terms for the same thing? Not just different uses of case like in the preceding example, but completely different words:

- Pen | Ink pen | Writing utensil
- Plug | Stopper
- Phone | Cell | That thing you use to talk to people with (!)

When you work in a very technical environment, engineers have different terms than UXers do. What they are out in the world and what they are in bits and bytes can be different, but everybody has to be on the same page as to what the connections and references are if that's the case.

I promised you a more content-focused taxonomy, didn't I? Okay, here you go:

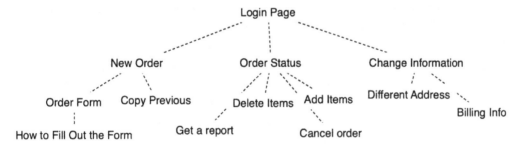

Figure 14.2 – Simple ordering application taxonomy

Figure 14.2 is a super simple illustration of an ordering platform, with three main choices to keep it easy, and just a couple of levels for them all. I urge you to copy this and rework it. See whether any of the subcategories would be better under a different main header. When working through these, a great idea is always a card sorting exercise. Have each member of the team responsible for content do it separately and see what you get. Some ideas may lead to consolidation of topics, and some may lead to new naming conventions.

The following are some additional examples of how taxonomies can be used to reduce the cost of managing content:

- A large retailer can use a taxonomy to organize its product catalog. This can help customers find the products they're looking for quickly and easily, and can also help the retailer to track sales data to identify trends.

- A government agency can use a taxonomy to organize its regulations and policies. This can help citizens find the information they need to comply with the law, and can also help the agency track changes to the law and make sure that its website is up to date.

- A university can use a taxonomy to organize its research papers and other scholarly materials. This can help students and researchers find the information they need, and can also help the university track its research output and identify areas of expertise.

As you can see, taxonomies can be used in a variety of ways to reduce the cost of managing content. If you're looking for ways to improve the efficiency and effectiveness of your content management process, then you should consider using taxonomies.

Exploring a few taxonomy tools

Depending on your current system's setup and/or needs, you can go dirt simple like I've done and use a tool such as **Scapple** (`https://www.literatureandlatte.com/scapple/overview`).

This will allow you to draft quick ideas for iteration and teamwork before you commit to using any of the more heavy-duty tools on the market. I won't go into each, but here's a short list for you to study and consider:

- **Coralogix Taxonomies**: This is a cloud-based taxonomy management tool that allows you to create, manage, and share taxonomies across your organization. It offers a variety of features, including the following:

  - A user-friendly interface

  - A powerful search engine

  - The ability to create custom taxonomies

  - The ability to share taxonomies with others

- **Sparx Systems Enterprise Architect**: This is a powerful modeling tool that includes a taxonomy management feature. It offers a variety of features, including the following:

  - The ability to create and manage taxonomies

  - The ability to import and export taxonomies

  - The ability to integrate with other systems

- **Taleo Taxonomy Manager**: This is a taxonomy management tool that is designed for use with Taleo's human capital management software. It offers a variety of features, including the following:

  - The ability to create and manage taxonomies

  - The ability to import and export taxonomies

  - The ability to integrate with Taleo's human capital management software

When choosing a taxonomy tool, it's important to consider your team and your company's specific needs and requirements. Some factors to consider include the following:

- The size of your organization

- The complexity of your content

- The features that you need

- Your budget

Once you've considered these factors, you can start to narrow down your options and choose the taxonomy tool that's right for you.

In addition to these commercial tools, there are also a number of open source taxonomy tools available on the market. These tools are often free to use and can be a good option if you are on a budget or you need a more flexible solution.

Here are a few of the most popular open source taxonomy tools:

- **Apache Jena**: This is a free and open source framework for building Semantic Web applications. It includes a taxonomy management tool that allows you to create, manage, and query taxonomies.

- **Fowler Taxonomies**: This is a free and open source taxonomy management tool that is written in Java. It offers a variety of features, including the following:

  - The ability to create and manage taxonomies

  - The ability to import and export taxonomies

  - The ability to integrate with other systems

- **Ontopia Ontology Manager**: This is a free and open source ontology management tool that includes a taxonomy management feature. It offers a variety of features, including the following:

  - The ability to create and manage taxonomies

  - The ability to import and export taxonomies

  - The ability to integrate with other systems

If you're looking for a free and open source taxonomy tool, then one of these options may be a good choice for you.

Now, let's make a smooth segue into strategies and tactics for making use of all the hard work you've been doing through holding interviews, writing and conducting surveys, and analyzing and synthing on all the metrics you've gathered from the tools you're using.

Utilizing your analytics work for content strategy

You have all the information from your work so now, how do you use it to actually create your content strategy? You have metrics, interview answers, maybe card sorting exercise findings, and so on. You've taken the time to create a taxonomy where none existed before, you've iteratively worked through the good and the not-so-good, the gaps and the holes, and now you're ready to go.

Summarizing your research work

This is the first step, because you want to get a concise picture of what the research has proven about your current content and content structure. Summarizing your research work is an essential initial step in the content strategy process, as it enables you to obtain a concise and comprehensive picture of the insights gained from your research efforts regarding your current content and content structure.

Through thorough research, including analytics, interviews, and card sorting exercises, you have gathered a wealth of information and data. Summarizing this research allows you to distill the key findings and takeaways into a condensed form, providing a clear understanding of the state of your current content.

By summarizing your research, you can identify patterns, trends, and common themes that emerge from the data. This summary acts as a valuable reference point, helping you to pinpoint the most critical areas of focus and inform the subsequent steps in developing your content strategy.

In this summary, highlight the key insights gained from each research component. Outline the significant metrics and analytics that shed light on user behavior, engagement, and content performance. Capture the main takeaways from interviews, including user preferences, pain points, and feedback. Additionally, summarize the outcomes of the card sorting exercises, highlighting any patterns or user-generated content structures that emerged.

Overall, summarizing your research work provides a solid foundation for developing an effective content strategy. It allows you to have a concise overview of the research findings and ensures that your strategy is based on data-driven insights rather than assumptions or guesswork. And it sets you up for the next step.

Writing a current-state content story

Now that your summary is ready, I want you to write out the current-state content landscape in a story. That's right: I want you to write it out, from your summary, into a story that anyone can pick up, read, and understand. You want a written account, and this will act as a content North Star of current-state content structure and strategy.

In my experience from years of work and hundreds of customers and use cases, this is a crucial step in leveraging the insights you've gathered. It allows the team and stakeholders to read through and get comfortable with the 'language' of the project. Keep in mind that this is the current state, not what you expect for the future.

In this content story, articulate the existing content structure, including its strengths and weaknesses. Outline the key findings from your analytics work, such as performance metrics, user feedback, and the outcomes of card sorting exercises. Describe the content gaps and holes you identified and how they're causing problems with users, and with the teams that manage the content if applicable. The key is to be unbiased and not interject feelings or preferences one way or the other. You must stay completely objective and as factual as your research work allows you to be.

In addition, the content story provides a foundation for developing a future content strategy. It sets the stage for strategic decision-making, guiding you in determining the necessary steps for content improvement and optimization. You can use the content story as a reference when identifying content priorities, developing content guidelines, and aligning content creation efforts with overall business objectives.

Remember, the content story acts as a powerful tool to communicate the current content landscape, establish a shared understanding, and lay the groundwork for shaping your future content strategy. By creating this written account, you can ensure that everyone involved is on the same page and working toward a common goal.

Summary

In conclusion, the process of leveraging analytics and research for content strategy is a vital component in creating impactful and effective content. By utilizing key metrics such as page views, time on page, bounce rate, and conversion rates, you can gain valuable insights into the success and performance of your content. These metrics provide indicators of content relevance, user engagement, and overall content strategy effectiveness.

Furthermore, the process involves summarizing your research findings to obtain a concise and comprehensive understanding of your current content landscape. By distilling the insights from analytics, interviews, and card-sorting exercises, you can identify patterns, strengths, weaknesses, and areas for improvement in your content and content structure.

With this knowledge in hand, you can then develop a content strategy that aligns with your target audience's preferences, addresses content gaps, and optimizes the user experience. This involves creating a content story that serves as a North Star, providing a clear overview of the current state of your content structure and strategy. The content story acts as a reference point for decision-making, content optimization, and prioritizing future content efforts.

And by leveraging analytics, research, and summarizing your findings, you can continuously refine and enhance your content strategy, making data-driven decisions to create content that resonates with your audience, engages them effectively, and achieves your business goals. With a solid understanding of your current content landscape and a strategic approach to content creation, you can build lasting relationships with your audience and drive meaningful results.

Next up: deliverables. You've done the work, found the problems, and have some ideas, so you need to show this to everyone involved. We'll learn the best practices for doing so in the next chapter.

Part 5: Tactics

This part of this book focuses on the essential aspect of delivering tangible outcomes and assets in UX Analytics projects. A successful project requires the delivery of findings at predetermined intervals throughout the engagement, as well as a final comprehensive delivery when wrapping up. These deliverables, agreed upon by the core team, serve as valuable assets and collateral for the project. They are not set in stone and can be adjusted based on the insights gained from analytics and other research as the iterative process unfolds.

Our brains are wired to process information visually, making visualization a powerful tool in delivering research and analytics findings. By presenting information in a visual form, we tap into the brain's natural preference for pictures, facilitating quicker understanding and decision-making. This chapter explores the importance of visualizing research and analytics and how it enhances the delivery of information.

Heuristic evaluations play a crucial role in assessing an application's alignment with human responses and behaviors. By employing heuristics, which are mental tricks that help us find shortcuts for common tasks and reduce cognitive load, practitioners can evaluate how well an application mirrors human responses to work. This chapter delves into heuristic evaluations and their significance in the context of delivering effective UX Analytics outcomes.

By understanding the importance of tailored deliveries, visualizing information effectively, and conducting heuristic evaluations, readers will be equipped to provide valuable assets and collateral that drive decision-making and enable the creation of exceptional user experiences.

This part has the following chapters:

- *Chapter 15, Deliverables - You Wrapped the Research, Now What Do You Present?*
- *Chapter 16, Data Visualization - The Power of Visuals to Help with Cognition and Decisions*
- *Chapter 17, Heuristics - How We Measure Application Usability*

15
Deliverables – You Wrapped the Research, Now What Do You Present?

Deliverables! Yes, the part that most UXers don't like very much. Okay, the part that most UXers hate! Let's be honest – we love doing the research, we love pouring through our analytics metrics, and we love finding the truth in the data. But after that, we just want to move on to the next project and start asking more questions.

But it doesn't have to be like that. With the right plan and iterative collateral creation, you can tame that beast. In the constantly evolving world of UX design, understanding the needs and desires of your users is paramount to creating exceptional digital products. UX analytics provides invaluable insights into user behavior, preferences, and pain points, guiding designers and developers toward crafting remarkable experiences. However, the true magic lies not only in collecting and analyzing the data but also in effectively delivering the findings to the teams responsible for making the needed enhancements and shaping the final product.

Welcome to the thrilling world of UX analytics deliverables! In this chapter, we'll embark on a journey that explores the pivotal role of reports, data, and collateral in shaping the success of your projects.

Here are the main topics of conversation:

- Why you must standardize deliverables
- The parts that can be flexible, but not compromised
- The research readout

So, buckle up as we uncover the secrets to creating compelling deliverables that will fuel your team's passion and propel them toward excellence.

Importance of standardizing deliverables

No UX analytics project is complete without findings being delivered. At predetermined times in a long engagement, and at the end when wrapped up, you will be delivering or giving access to assets and collateral for the team. These deliverables will be agreed on by the core team and can be adjusted based on analytics and other research as the iterative process moves through.

Let's imagine for a second that three people are responsible for creating and delivering reports via spreadsheets, PowerPoint decks, PDF images of metrics, and maybe dashboards that stakeholders have access to. With no deliverables standards on fonts, one could use Arial, one could use Times New Roman, and the other could use Helvetica. This is a simple example, I know, but think about your reaction if you dip into a folder and pull up documents that are all over the place with fonts. You're going to think that the UX team is unprofessional and that's not the reaction we're aiming for!

And do you report everything via text documents, or do you use presentation decks? Do you show all quant data in spreadsheets that you're using as a form of presentation? With red blocks and yellow blocks and green blocks. With dark fonts on dark backgrounds. You've seen these I know and if you're like me you cringe.

If the company you're doing work for puts out reports like this, you've got some work to do. Standardizing deliverables is not just a best practice for the sake of a nice and tidy look and feel – it also strengthens recall. When you have a visual system of delivery – fonts, colors, charts, and so on – and every form of your collateral adheres to these standards, you allow those consuming your information to recall previous information faster and with greater accuracy. So, keeping all of this in mind, let's get down to it.

The art and science of designing insightful standardized reports

The way we talk about "reports" in this context is very broad and meant to mean anything you design and produce to be shared with others. Think back to what you learned about creating the North Star statement and how it's a collaborative piece of work that's understood and agreed upon by the stakeholders and your team as representatives of the work that's expected to be completed. That work is similar to this work in that they both articulate the work to be done.

Learning how to transform raw data into narratives that captivate your audience and ignite their imagination is an important step for all Uxers and all analytics teams. So, to foster that experience, we'll delve into techniques for structuring reports, visualizing data, and presenting key findings in a way that drives action and inspires the solution discovery team.

In the realm of UX analytics, data is the raw material that holds immense potential. However, it's through the art of storytelling that we unlock the true power of this data and create reports that captivate our audience, igniting their imagination and driving them toward action for change.

A captivating story can transcend mere numbers and graphs, transforming data into a narrative that resonates with stakeholders and teams alike. It breathes life into the insights uncovered, painting a vivid picture of user experiences, challenges, and aspirations. By mastering the art of storytelling within your reports, you can shape the direction of projects, evoke empathy, and inspire creative solutions.

But how do we craft such reports that leave a lasting impact? How do we structure them in a way that grabs attention and compels action? These are great questions. Let's dive into a few techniques to help you transform your reports from mundane and lackluster data repositories to captivating narratives that drive your solution discovery team forward.

Embracing the power of structure

Just like a compelling book, your report design should have a well-defined structure that guides your audience through a coherent and engaging journey. A well-defined structure not only provides clarity but also instills a sense of anticipation and purpose. Begin with a captivating introduction that sets the stage, highlighting the significance of the research and its potential impact. Though you should be presenting this report in a read-out to the team, also consider that it should all stand on its own. Anyone should be able to get to the folder, pick it out – and any other supporting collateral – and run with it.

> **Note**
>
> In many cases, those people who you present to, including the solution discovery team, will need to report your findings up the food chain to the next highest level of management. When you consider that the audience for your work is much broader than just the stakeholder, you'll understand the importance of the need for structure and consistent style, according to the standards set.

As you progress through the body of the report, skillfully weave together key findings, supporting evidence, and analysis logically and coherently. Each section should seamlessly flow into the next, creating a narrative thread that keeps readers immersed in the story you are telling.

Finally, when you conclude, leave your audience with a powerful call to action that empowers and inspires your team to utilize the insights gained from the report to shape exceptional user experiences. By structuring your report thoughtfully, you ensure that every page becomes a stepping-stone toward achieving your project's goals and delivering remarkable outcomes.

Considering the bigger picture

Though this should be in one form, say a presentation deck, all of your supporting evidence and collateral will be in various forms and stored in the folders within your repository. This requires a tight naming convention – sounds like taxonomy work, doesn't it? – so that anyone can find what they need as the project progresses.

How about an example?! I hear you and present the following:

Folder structure

Team or division – (retail, corporate, customer service, and so on)

Project name:

- Research categories:

  - Data, metrics - data

  - Interviews - int

  - Design thinking - dt

    - Process maps - pm

    - Journey maps - jm

  - Reports and collateral for consumption:

  - Decks or PDFs - deck

  - Documents - doc

  - Data Visualization | Charts - dv

Naming convention:

Research: project code_category_date_name:

- `B17_int_0825_shippingmngr1.docx`

- `B17_data_0826_acctsrecv.xls`

Reports and collateral for consumption: project code_category_date_name:

- `R40_deck_0826_finalreadout.pptx`

- `R40_dv_0828_salesMetricsNortheast.jpg`

In essence, always consider the bigger picture. Always think upstream and downstream. Who might need to look at your research collateral in HQ, and who might need to refer to your work in stores, plants, hospitals, dealerships, and so on? The work you do is monumental and should serve many purposes within the organization.

To take this a step further, think of all this as the documentation you need to prove your work. At times, you'll need to cite research done to prove the legitimacy and accuracy of your numbers. All of the work you've done should be easily accessible so that you can refer to any document or any dataset with ease. Never leave it to chance and never assume that everyone will take you at your word – some people require proof at a very granular level.

In addition to citing research, documentation serves as a comprehensive record of your UX analytics process, methods, and results. By maintaining organized and easily accessible documentation, you can save time and effort when referring to past research, supporting your arguments, or making informed decisions based on previous findings.

Furthermore, documentation not only serves as a reference for your work but also facilitates collaboration and knowledge sharing within the organization. By making your documentation accessible to relevant stakeholders, such as designers, developers, and other members of the solution discovery team, you promote a shared understanding of the research and encourage cross-functional collaboration. This enables team members to access the necessary information, validate your findings, and contribute their expertise to the decision-making process.

It's also important to realize that different stakeholders may have varying levels of familiarity or comfort with the subject matter. Therefore, thorough documentation becomes crucial in addressing such needs. It allows you to present evidence at a granular level, supporting your assertions with precise data points, research methodologies, and references. This level of documentation helps establish credibility, builds trust, and ensures that your work is grounded in robust evidence.

Researching level metadata

This will be documentation about your documentation, so to speak. As mentioned earlier, you need to be able to talk about your methodologies. Have a small collection of documents – whatever is required – that carry all direct references to the way you did the research. Document the metrics derived from the tools that are watching clicks and pathways through your applications. Document the research plan so that people can understand your reasoning.

Write out anything that refers back to previous work. If, for instance, you and the team did work on another project that has a similar scope to this, and the user base is similar, use the parts of that research that fit and point that out. Essentially, don't leave anything to chance. Expect that you'll have to prove every decision you and the team made. Record conversations, virtual team calls, and any time you're presenting at milestones during the project. Always make note of opposition or setbacks that occur, and always make note of blockers that slow progress or bring it to a crashing halt.

Visualizing for impact

Data visualization is an indispensable tool in the realm of UX analytics (we'll cover this subject in depth in the next chapter) because it enables you to transform complex information into a visually compelling narrative. By incorporating charts, graphs, and infographics, you provide your audience with a visual roadmap that enhances comprehension and retention. But it goes beyond just conveying information – data visualization has the power to evoke emotions and create a lasting impact. Strategically utilizing colors, typography, and imagery allows you to craft a visual language that aligns seamlessly with your narrative, amplifying the story you're telling.

By engaging both the analytical and creative sides of your audience, you create an immersive experience that enables them to absorb information more effectively and form a deeper connection with the insights presented (remember the thinking types you learned about earlier). Visuals not only facilitate understanding but also spark curiosity and inspire innovative thinking, setting the stage for collaborative exploration and a shared drive toward designing exceptional user experiences. With data visualization as your ally, you unlock the potential to transform raw data into a compelling visual story that resonates with your team and drives actionable outcomes.

The next chapter is on data visualization, and we'll get into data representation and chart types there, so this is a good precursor to that. Right now, however, just internalize that data visualization is a tool you can use to create impact in your presentations.

Telling the user's story

When creating reports for a diverse audience comprising designers, developers, and stakeholders, it is essential to maintain a user-centric focus. By incorporating real-life anecdotes, quotes, or testimonials, you can humanize the data and bring the user's experiences to life. Introducing personas or user journeys allows readers or viewers to empathize and understand the users' needs and pain points by stepping into their shoes. Placing the user at the center of your narrative creates a sense of purpose and urgency that motivates action – this is empathy building. Sharing the stories of actual users and their encounters with the product or service fosters relatability and emotional engagement, offering stakeholders and team members valuable insights into the lived experiences of users.

To introduce personas or user journeys, you can begin by creating fictional characters that represent the different user segments. These personas allow your audience to step into the shoes of these individuals. Personas breathe life into data points, turning them into relatable personalities with specific needs, desires, and pain points. User journeys, on the other hand, provide a visual representation of the user's interaction with the product or service over time. This holistic perspective fosters empathy and deepens the understanding of the user's journey, enabling stakeholders and team members to grasp the context in which decisions are made.

By placing the user at the center of your narrative, you imbue your reports with a sense of purpose and urgency. It reminds your audience of the real people who'll be impacted by their design and development decisions. This user-centric focus creates a shared commitment to delivering exceptional user experiences and drives action. It motivates your audience to seek innovative solutions that address the needs and pain points that were uncovered during the UX analytics process.

Maintaining a steadfast and non-biased focus on the end user throughout your report is paramount. By infusing real-life stories, leveraging personas, and exploring user journeys, you can foster empathy and deepen understanding. Placing the user at the heart of your narrative generates a sense of purpose and urgency, driving your audience toward action and inspiring them to create remarkable experiences that resonate with the needs and aspirations of those who will ultimately use the product or service.

Be clear, concise, and actionable

As you embark on crafting your report, clarity and conciseness should be your guiding principles. You must strike a balance between providing sufficient information and overwhelming your audience with unnecessary details that could obscure the main message. Your goal is to deliver a report that is focused, digestible, and impactful. Again, to strengthen what we mentioned earlier, they must ensure that they design and build the best new product or make the best enhancements to the existing product.

To achieve clarity, distill complex concepts and findings into concise, easily understandable language. Avoid technical jargon or industry-specific terminology that may alienate or confuse your audience. Rather, strive for a clear and accessible writing style that ensures everyone can grasp the core ideas and insights being presented.

Remember, the purpose of your report is to guide the solution discovery team toward actionable outcomes. So, to best enable that, emphasize the key insights that have the potential to drive meaningful change. Identify the most crucial findings that have practical implications and prioritize them in your report. By doing so, you provide your team with a clear roadmap and actionable recommendations that lead to tangible improvements in the user experience.

While focusing on clarity and conciseness, it's important to maintain a compelling narrative. Each word in your report should contribute to the overall story you are telling. Craft your language carefully to engage and captivate your audience. Choose words and phrases that evoke emotions, spark curiosity, and inspire action. By employing a persuasive writing style, you can ensure that your report not only informs but also motivates the solution discovery team to act upon the insights presented.

Additionally, consider the structure and formatting of your report to enhance clarity. Use headings, subheadings, and bullet points to organize information in a logical and easy-to-follow manner. Visual elements such as tables, diagrams, or illustrations can also aid comprehension and break up dense blocks of text.

A word about visualizations

Data visualization plays a pivotal role in complementing the clarity and conciseness of your report. While crafting a clear and impactful narrative, it's important to leverage data visualization techniques to transform complex information into visually compelling representations. Visualizations provide a powerful means of communicating patterns, trends, and relationships within the data, enabling your audience to grasp the information quickly and intuitively. They serve as visual anchors, highlighting key findings and supporting the narrative you have constructed.

Effective data visualization not only engages the analytical side of your audience's minds but also taps into their creative faculties, fostering a deeper connection with the insights presented. By harnessing the power of visuals, you elevate your report to a higher level, where information becomes engaging, memorable, and influential, leading to more informed decision-making and the creation of impactful UXs.

The power of storytelling within UX analytics reports that combine effective writing with supporting visual elements cannot be overstated. By transforming raw data into compelling narratives, you can captivate your audience, ignite their imagination, and inspire them to take action. Through effective structuring, compelling visualization, user-centric storytelling, and clear communication, you can create reports that not only inform but also motivate your solution discovery team to create exceptional user experiences. So, embark on this journey of storytelling and unlock the transformative potential of your data.

Coming back to structure and standards

You just learned a lot, but what's important to remember is structure and standards. Each part of your report should complement the others. If you use Arial in the text copy, Arial should also be used in the titles and tags of your visualizations. If you have branded icons, colors, or anything else, make sure you use those elements.

Another consideration is that your work can – and should – become part of a much larger plan for the product. Often, companies have playbooks that are used by production and engineering, and the inclusion of your UX analytics work can help strengthen the narrative that aligns all facets of the company. Continue to expand your understanding that UX is part of the larger picture and that your work will start to support it all.

Delving into parts that can be flexible but not compromised

In the realm of UX, just like in life, things don't always go according to plan. Sometimes, circumstances demand flexibility, requiring us to adjust, pivot, or even abandon certain approaches. Tight deadlines may force us to skip steps and adapt our delivery process. However, learning to be flexible in the delivery of your UX analytics research is a fundamental aspect of becoming a seasoned UX professional. It involves recognizing when it's acceptable to adjust without compromising integrity or thoroughness while navigating the daily challenges faced by every company in the world. Flexibility within the delivery of your work allows you to strike a balance between agility and quality, ensuring that you can adapt to the ever-changing demands of the UX landscape.

One key aspect of maintaining flexibility in your UX analytics deliverables is understanding the core objectives and priorities of each project. By identifying the critical elements that must be preserved and the ones that can be adjusted, you can make informed decisions about where to be flexible without compromising the overall quality and value of your work. This requires clear communication with stakeholders and a thorough understanding of their needs and expectations, as we learned in previous chapters. By aligning your flexibility with the project's goals, you can navigate the challenges effectively while still delivering valuable insights.

In situations where you and your team had initially planned a second round of interviews or surveys to gain clarity on certain aspects that were not well-defined, there may be instances where you are asked to expedite the process and provide immediate findings. In such cases, it is helpful to infer the insights that you and your team can gather, while also clearly indicating in the deliverables that further investigation would provide a more comprehensive and precise understanding. By acknowledging the need for additional exploration, you ensure transparency and provide a roadmap for obtaining a clearer picture in the future.

Example

If, for instance, your research up to this point has shown that 15% of your corporate buyers have trouble with finding their previous orders, and you feel that an additional 20 minutes of inquiry with 10 of them would give you the information needed to provide evidence that a simplification of the process would cut that down to 5% within 2 weeks, then you should state that with your readout.

Furthermore, flexibility in delivery also involves being resourceful and adaptable in your approach. Sometimes, constraints, such as limited resources or time, necessitate finding creative solutions to achieve the desired outcomes. This might involve streamlining processes, leveraging existing data sources, or employing innovative techniques to gather insights more efficiently. By embracing a flexible mindset, you can explore alternative paths, optimize your workflow, and overcome obstacles without sacrificing the integrity of your research. It's about finding the balance between being adaptable and ensuring the robustness of your methods.

I must restate this – flexibility should not be confused with compromising quality or taking shortcuts. While you may be flexible in certain aspects of your work, such as adjusting timelines or methodologies, it's crucial to maintain the utmost professionalism and uphold the standards of UX research. This means adhering to ethical guidelines, ensuring data accuracy, and delivering thorough analysis and actionable recommendations. By being selective in where flexibility is applied, you can navigate the challenges of UX analytics while upholding the integrity and value of your deliverables.

Flexibility in the delivery of UX analytics research is a crucial skill for seasoned UX professionals. By understanding the project objectives stated in the North Star statement and agreed upon, prioritizing core elements, and maintaining a resourceful and adaptable approach, you can strike a balance between agility and quality. It's important to be mindful of the boundaries and maintain professionalism, ensuring that flexibility does not compromise the integrity and thoroughness of your work. Therefore, by embracing flexibility in the right areas, you can navigate challenges, meet deadlines, and deliver valuable insights that drive exceptional user experiences.

Working from templates

Interview blueprints are templates. If you're in-person, you print them out and hand a few to each of your core team members. If you're virtual, just make sure that everyone who needs to take notes has a copy and is ready to go.

Your text documents should all be derived from a standard template. They should have any heading that's needed and the correct fonts and font weights, colors, and so on. If you're using PowerPoint or any other software to create decks, those should all be created from standard templates. The company branding is usually built into these, so be sure to use them and show your work in a visual style that anyone in the company will recognize immediately.

Remember that you want people to absorb your information naturally. You don't want to cause them to struggle to think or understand, and the more distractions you take away, the easier it becomes for them. Hence – templates. Start using these if you don't and make your life and the lives of all your team members easier.

Mastering the research readout

The terminology for this practice may vary across companies, so if the specific term is unfamiliar to you, there's no need to worry. It simply refers to the act of presenting and explaining the findings of your unbiased research on the problems that need to be addressed. With all your research completed and your deliverables designed using the principles discussed earlier, it's important to ensure that your research readout follows a standardized approach.

This consistency allows your audiences to know what to expect and facilitates a smooth transition from problem discovery to solution discovery. The research readout serves as a formal aspect of the product design life cycle, providing a structured and transparent handoff of insights that will inform the subsequent stages of the UX process.

When preparing for the research readout, it's important to keep the diverse range of stakeholders who will be present in mind. Tailor your presentation to accommodate varying levels of technical knowledge and familiarity with UX concepts. Clearly communicate the purpose of the research readout, emphasizing its value in informing the decision-making process and driving the development of user-centered solutions. Setting the stage and managing expectations ensures that your audience understands the significance of the insights you will be sharing.

During the research readout, strike a balance between providing a comprehensive overview and avoiding overwhelming your audience with excessive details. Focus on the key findings and insights that directly relate to the problems that were identified during the research phase. If you can work in a confidence rating, do so. It can help your listeners solidify thoughts in their heads while assessing possible work to be done and money to be spent. Use visual aids, such as charts, graphs, and examples, to illustrate the data and make it more digestible. Additionally, make sure you address any questions or concerns raised by the stakeholders, fostering a collaborative environment that encourages dialogue and a deeper understanding of the research findings.

As part of the research readout, emphasize the unbiased nature of your research. Clearly articulate the methodologies that have been employed, highlighting the measures taken to ensure objectivity and minimize biases. By transparently explaining the research process, you build trust and credibility with your audience, reinforcing the importance of data-driven decision-making. In tandem, you're enabling stakeholders to make informed decisions and drive the development of user-centric solutions that address the identified problems.

Tone and eye contact

As of this writing in 2023, we've lived through a global pandemic that's changed the way we work – I'd wager forever. We're getting more and more used to seeing our colleagues on our laptop screens as remote work has become a normal way of life for most of us. I'll cover both virtual and in-person presentations here, but I especially want to emphasize the most important aspects of virtual presentation.

Developing tone

Similar to tone when interviewing, the way you verbally deliver your findings is important. It will also depend on your audience and the expectations that have become normal when UX interacts with other teams and business units. If you've evangelized UX throughout the company and have created allies up and down the corporate structure, there will be one style. If you're a consultant working on a project within a company you've been hired out to, then the approach will be a bit different. The key is knowing your material as much as you know your audience.

Always be professional but understand that your audience is keying on your delivery as much as what you're showing in your deck, or whatever you've decided on. Most likely that's a deck but hey, you may have figured out a more engaging way. If you show enthusiasm for your findings, your audience will pick up on that, so don't be afraid to show some excitement. With multiple monitors being used by most of us now, some will have your presentation on one screen and the virtual team screen on the other so that they can look back and forth between the video of you and the deck. If you're smiling and have a confident look on your face, that sneaks through the ether and creates a strong bond between you and the viewer/listener.

> **Special note**
>
> I doubly encourage you to listen to recordings of your presentations and count the number of "um"s you use per minute. This isn't an easy habit to kick for sure, but training yourself to just pause rather than uttering that pseudoword is crucial. I also strongly believe that if you know your stuff, you won't be in the position of filling in blank spaces with an "um" or "uh." I coached a very brilliant researcher on this recently and he thanked me for bringing this up. We walked through ideas and I asked him to practice. Within the next few presentations he did, he noticeably reduced that dreaded utterance considerably. If this is a big thing for you, I know you can correct it with awareness, diligence, and practice.

Eye contact

If you take away nothing else from this section, I want you to solidify this part. I mean get this, internalize it, and master it.

LOOK. INTO. THE. CAMERA.

Let me repeat that one more time: Look into the camera. I can repeat that again if you need it. No? Okay.

When you're presenting, I know you have to look at your deck and material. Hopefully, you're not reading every word in the text – which should be minimal – but yes, I know you must look at your screen. But if you know your stuff, look at the screen to get where you are, then into the camera to continue speaking. This is like the way an on-air news person works. They glance at their notes and then into the camera.

Why? Because people are looking at you! And if all you're doing is looking at your screen and reading all the words, you aren't looking at them – and you MUST look at them. Think about this. You're sitting across the table from an important stakeholder, discussing the project. Are you going to look down at your notes or laptop the entire time? No! You're making eye contact and having a conversation with this person, who by the way, is very important to you.

The same holds for when you're doing any on-video meeting. Look into the camera, not at the video or yourself or anyone else. When you do this, they all see you looking down at your screen, and not at them. Please practice this and make it a habit. Your presentations will be much better because of it.

Summary

Without well-thought-out and well-designed deliverables, all your work is for nothing. In this chapter, we covered essential topics related to effectively presenting research findings and insights. With an emphasis on standardizing deliverables to ensure consistency and clarity in communication, we learned about structure and consistency. While certain parts of the deliverables can be flexible to accommodate specific project requirements, compromising on quality or essential components is discouraged.

This chapter also explored the significance of the research readout, which involves presenting research findings comprehensively and engagingly. By following these guidelines, you and your team can create impactful deliverables that effectively communicate research insights and drive informed decision-making.

In the next chapter, we'll cover data visualization. We will delve into the power of data visualization in enhancing cognition and facilitating informed decision-making. Through effective visual representation of data, complex information can be simplified and understood more easily, unlocking valuable insights and empowering users to make data-driven decisions.

Data Visualization – the Power of Visuals to Help with Cognition and Decisions

Data visualization is one of my favorite topics, and the study and integration of it into my work output over the past 20 years has provided great value to many companies and agencies across the world. This accounts for thousands of people who, until they were allowed to discover the power of data visualization, were struggling through stacks of spreadsheet printouts and disjointed information sources.

It's powerful and impactful because our brains work in pictures, not words or numbers. By visualizing research and analytics, you can deliver information in a form that allows for quicker understanding and decision-making. The transformative power of data visualization lies at the heart of this process, enabling us to unlock hidden patterns, derive actionable intelligence, and make informed decisions that drive businesses forward.

In this chapter, we'll get very familiar with the captivating world of data visualization, exploring its diverse facets and its pivotal role in the realm of UX analytics. Here are the main topics of conversation:

- Data visualization for quick insights
- Static versus interactive visualizations
- How we move toward predictive and prescriptive analysis

Data visualization for quick insights

Data visualization acts as a gateway, bridging the gap between raw data and human understanding. Transforming complex datasets into visually compelling representations empowers both analysts and stakeholders to grasp the underlying narratives and uncover valuable insights that might otherwise remain concealed within the labyrinth of numbers and figures.

> **Note**
>
> It's a language that speaks to our visual brains, effortlessly conveying information, triggering cognitive connections, and facilitating comprehension.

Today's world never stops, never sleeps, and constantly demands more and more of us if we're not working smartly and efficiently. Time is a prized commodity, so the ability to swiftly extract key insights from data is paramount.

Within this section, we'll explore the power of data visualization as a tool for gaining quick insights. We'll uncover techniques and methodologies that allow us to distill complex information into concise, visually compelling displays. By mastering the art of data visualization for quick insights, you'll learn how to identify trends, spot anomalies, and derive actionable intelligence efficiently, setting the stage for timely decision-making.

Understanding basic data visualization types

Although I'm quite sure you're familiar with these types of charts, we're going to start here and build upon the information as we get more complex and granular. So, we'll start at 30,000 feet and then dive down into the woods – so to speak. We're going to work with bar charts, donut charts, line charts, and scatter plots as our starting point – all with visual examples so that you start grasping the concepts immediately. Because – say it with me – we think in pictures. So, it kind of makes sense to learn about data visualization with data visualization. Do you agree?

Bar charts

Among the many varied data visualization techniques, bar charts are undoubtedly the most recognized, offering a visually intuitive and impactful means of conveying information. With their clean lines, clear comparisons, and straightforward representations, bar charts have long been a favored tool in the arsenals of analysts and storytellers alike.

One of the key strengths of bar charts lies in their ability to facilitate clear comparisons and rankings. By visually representing data points as varying bar lengths, the human eye can quickly discern disparities and patterns within a dataset.

Take, for example, this illustration of a side-by-side column chart in which we start to compare sales in five apparel categories, men's versus women's:

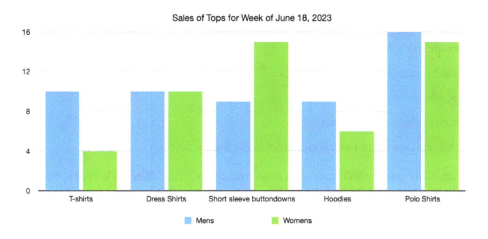

Table 1

	Mens	Womens	
T-shirts	10	4	
Dress Shirts	10	10	
Short sleeve buttondowns	9	15	
Hoodies	9	6	
Polo Shirts	16	15	

Figure 16.1 – Simple 2x5 category side-by-side column chart

Or, this version, a stacked column chart:

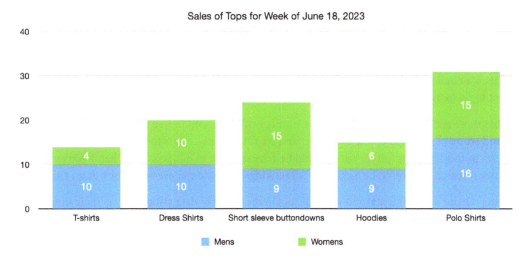

Figure 16.2 – Simple 2x5 category stacked column chart with value annotations

Let's consider immediate reactions possible when someone looks at these charts:

- There are two colors – men's and women's
- Polo shirts outsold all other categories
- There are five categories
- Women's t-shirts sold the least
- The biggest gaps are in t-shirts and short-sleeve button downs

The first, immediate reaction will most often be the two colors. That difference is what jumps out in a split second. Of course, considering 508 compliance, people with different forms of color blindness will see this differently, but they will see the variance in shade and be able to distinguish between the two.

---

508 compliance

508 compliance, also known as Section 508 compliance, refers to adherence to the accessibility standards outlined in Section 508 of the Rehabilitation Act. Section 508 requires federal agencies to ensure that their **electronic and information technology (EIT)** is accessible to individuals with disabilities. It sets guidelines and criteria for accessibility, including the provision of alternative text for images, compatibility with assistive technologies, and the availability of accessible features for individuals with visual, hearing, or mobility impairments. Compliance with Section 508 aims to ensure equal access and usability of digital content and technology for all individuals, regardless of their abilities.

Next might be the variances in bar sizes. The eye might go to the two tallest bars, and then to the label to figure out what's selling the most. Or, it could go to the shortest bar – women's t-shirts – to recognize what's a slow seller.

We'll get into what people can do with these facts a bit later, but for now, realize that this visual will allow someone to do something – to make some sort of a decision based on sales goals, most likely, since this is a retail store. If the women's t-shirt sales are troubling, the store manager needs to figure out what to do to increase those sales.

Note that these are charts of quantities – the number of items sold in five categories, split between men's and women's styles. A count was done, and we totaled the counts for each. You'll tend to use bar charts in this way most often, but it would be remiss of me to not point out that if you're doing qualitative research, this could be a chart of 1-5 rankings on customer service scores, as an example.

Bar charts remain a fundamental and versatile tool in the world of data visualization. Their simplicity and clarity make them an ideal choice for unveiling insights with visual precision. By understanding the nuances of bar chart design, leveraging their capabilities for comparison and ranking, exploring variations, and enhancing them with annotations and interactivity, you can harness the full potential of this timeless visualization technique.

Donut charts

In the realm of data visualization, donut charts emerge as a simple and recognizable, yet dynamic, visual tool offering a unique and engaging way to show proportions and relationships within a dataset. With their circular structure and clever use of arcs and slices, donut charts provide a visually pleasing representation that enables viewers to grasp the distribution of data at a glance.

In my practice, donut charts have replaced pie charts. They're easier for most people to understand, and the white space at the center separates the sections better. They're best suited to understanding parts of a whole, most often as percentages. Let's take a look:

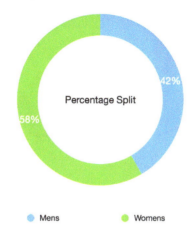

● Mens   ● Womens

Table 2

	Mens	Womens		
T-shirts	5	7		
Dress shirts	10	8		
Short sleeve button downs	9	15		
Hoodies	7	12		
Polo shirts	16	21		

Figure 16.3 – Simple donut chart showing split of two main categories

Okay, so what are the immediate reactions here?

- Two colors
- The green part takes up more space than the blue
- The two add up to 100%
- It doesn't include the items
- Women's is outselling men's

So, again, the first reaction will most likely be to the two colors. I think you can agree, however, that in this donut chart, that's more evident than in the bar chart because they're only depicted once each. The next immediate reaction – and that's in, like, a split second – is that the green occupies more space than the blue.

Because of the simple form of this illustration, it doesn't include the item type breakdowns. That can be done like this:

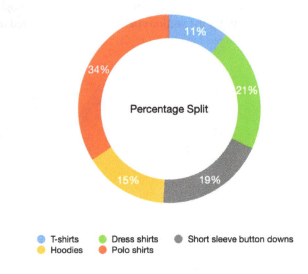

Figure 16.4 – Different categorization of donut

Now we can see the more granular breakdown of the different apparel items (where we did very well with polo shirts!) and a great way to visualize this is to put the two charts side by side:

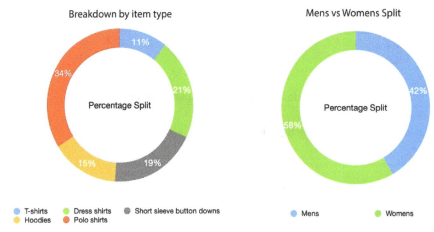

Figure 16.5 – Comparison of two data views

Now we're starting to let the data tell a story that's more relevant to a manager or an analyst. If we imagine that the scope of this information covers sales in 20 large department stores in the United States, we can see that it will give a marketing or sales manager some quick info on which to adjust if needed.

Another thing I like donut charts for is that they also give the viewer a quick idea of how many factors are involved in a dataset. In *Figure 16.5*, we have two on one chart and five on the other. When someone is new to a dataset especially, they can quickly see what they're dealing with. When there are many sections in a donut, they know there's a lot to sort out. If one of those is much larger than the others, as in this example, they'll have questions.

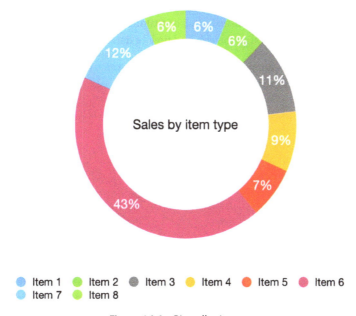

Figure 16.6 – Big seller impact

We see very quickly that we've got one item type that's far outselling the others. If we planned for this to happen, the donut shows us that we're successful. If we didn't, the donut shows us that either we have a problem, or we have a good thing happening that we want to capitalize on. One way or the other, this donut chart shows us actionable data that we can further analyze and make decisions from.

Donut charts, with their circular allure and ability to communicate proportions, offer a distinct perspective in the realm of data visualization. By understanding their strengths and appropriate usage, and carefully designing for clarity, you can harness the potential of donut charts to unlock a ring of insights.

To ensure effective communication, careful design choices must be made when creating donut charts. Labels should be positioned strategically to avoid clutter and ensure clarity, guiding viewers' understanding of each slice's significance. Thoughtful color selection can aid in distinguishing between categories and highlighting key insights.

Line charts

With their simplicity and elegance, line charts offer a visual language that effortlessly communicates the evolution of quantitative data over time. By plotting data points along a continuous axis, line charts provide a coherent visual representation of how values change over time. This temporal analysis enables analysts to discern long-term trends, fluctuations, and seasonality within data. Whether you're examining stock market performance, website traffic, or customer satisfaction ratings, line charts serve as a reliable compass, guiding us through the ebb and flow of data dynamics.

Here's an example of sales figures for stores in the north region versus stores in the south region, by day over an imaginary one-week period.

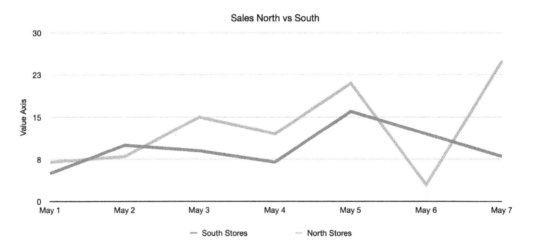

Figure 16.7 – Simple line chart comparing temporal data

A line chart can be as granular as a minute for some applications, or as broad as a year. In most charting software, you can choose hour, day, week, month, quarter, or year. If it's in your data, you can visualize and analyze it.

In the example shown in *Figure 16.7*, our user will immediately notice the huge dip on **May 6** for the north stores, followed by a big jump back up. It's out of character with the sales trend for the other days so somebody needs to check to see what happened. Let's imagine some possible causes. I'm going to list a few, and then I encourage you to come up with some of your own hypotheses:

- Data error
- Weather-related
- Power outage in town or store
- Shipping delay of product

Now see how many more you can add to the list. What else can you imagine could have caused a large drop in sales like that? Remember this is for the entire north region, not just one store. See if you can add 5-6 more ideas.

To effectively communicate data trends, line charts require careful design considerations. Clear labeling of axes, including appropriate units and time intervals, ensures that viewers can easily interpret the data. Thoughtful scaling of the axes allows for accurate representation and optimal use of the chart's visual space.

Also consider the visual styling of the lines, including color choices and line thickness, to guide the viewer's attention and emphasize specific trends or data series. By focusing on these design elements, line charts can transform complex time-based datasets into visually accessible narratives that invite exploration and analysis.

Scatter plots

With their unique ability to represent two continuous variables simultaneously, scatter plots provide a canvas where data points scatter, allowing us to navigate the multidimensional landscape of our data. By plotting data points on a Cartesian plane, where one variable is represented on the x axis and the other on the y axis, scatter plots provide a visual representation of how the two variables interact. This mapping enables analysts to identify correlations, clusters, outliers, and trends, fostering a deeper understanding of the complex dynamics at play within the dataset.

We'll continue to use our retail sales examples for this one, and we'll see if we can figure out why that dip occurred on May 6. If we keep in mind that retail sales are dependent on a variety of factors, we can hypothesize. Ready?

I think we should check the weather in the north region for the week depicted and see if that has any bearing. My team and I found a database of weather for the region and plugged that into a scatter plot to compare against our sales numbers.

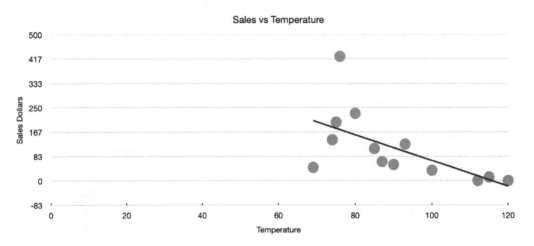

Figure 16.8 – Scatter plot of sales versus temperature with trend line

So, what we discover is that there was a massive heat wave that saturated the north with intense heat and humidity. You notice that as the temperature rose rapidly, sales plummeted, which gives us an inverse correlation. As one metric goes up, the other goes down. What are some of the things we can infer at this point?

- It was so hot nobody wanted to go outside
- It was so hot there was a heat advisory and everybody listened
- It was so hot that power outages hit the entire region

What more can you think of?

While most of the sales figures are close, there's one outlier of approximately $420 before the temperature got extreme, but that doesn't change the fact that the intense heat killed our sales.

Let's look at another example, of cash versus net income, and this time with more data points and sizing the bubbles:

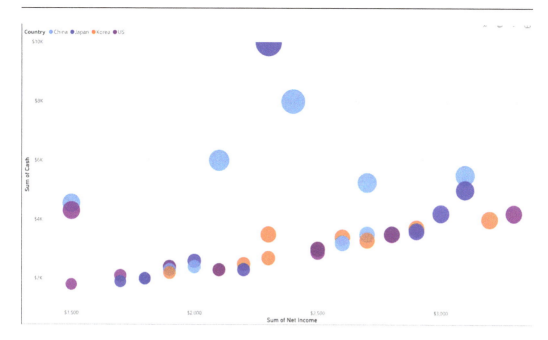

Figure 16.9 – Another version of a scatter plot, with sized bubbles

In this scenario, we can imagine that each bubble is a store and the cash is money received via retail and online sales, with the net income being the cash minus operating expenses. You'll notice a fairly similar trend with most of the data points following a linear path, but then also a few larger bubbles that are outliers compared to the majority of the other data points.

Sizing is done on this chart by cash, and the chart shows us very clearly that although cash was high – vertical axis – the net income was low to middle on the chart – horizontal axis – which indicates that there's a problem with these stores. They had a lot of cash intake, but after expenses, they didn't have the correct ratio of net income. So, this chart very quickly would let a regional manager perhaps see problem areas and do what's required to task the UX team to investigate via analytics to pinpoint the problem.

So, as you can see through the two examples given, scatter plots offer a unique vantage point to uncover patterns within data that may not be immediately apparent in other visualizations. Clusters of data points may indicate subgroups or segments within the dataset, providing valuable insights into customer segments, market niches, or distinct patterns of behavior. Outliers, represented by data points that deviate significantly from the overall trend, can shed light on anomalies, exceptional cases, or points of interest that warrant further investigation. Additionally, the visual examination of scatter plot trends, such as linear or nonlinear relationships, can uncover underlying dynamics and guide future analysis or modeling efforts. Through the lens of scatter plots, patterns emerge, connections are formed, and new avenues for exploration open up.

Static versus interactive visualizations

Being both a matter of choice and context, static and interactive visualizations are both powerful ways to help your user base understand how your UX analytics work. We're going to explore the pros and cons of static and interactive visualizations in this section so that you can make the best decisions for your purposes.

Static visualizations excel at succinctly summarizing information, enabling us to present key findings in reports and presentations. We'll refer to these visualizations as snapshots. On the other hand, interactive visualizations empower users to explore data at a granular level, facilitating deep analysis and discovery. By understanding the strengths and weaknesses of each approach, you will gain the flexibility to employ the most effective visualization techniques for your specific analytical and presentation needs.

Static visualizations

What's shown in the earlier examples are static visualizations. They're pre-generated images or charts that communicate information in a fixed format. They offer a snapshot of data at a particular moment and lack user interactivity. Static visualizations are widely used for quick reporting, presentations, and sharing insights through various mediums such as reports, slides, or printed materials. They provide a concise and straightforward representation of data, allowing users to grasp key information efficiently.

The benefits of static visualizations lie in their simplicity and ease of understanding. They are particularly useful when the audience requires a quick overview or when the data itself doesn't require further exploration. They're well suited for providing high-level views and enabling managers and executives to quickly grasp the overall state of their organization. These visualizations offer concise and easily understandable representations of key metrics, trends, and comparisons, allowing directors to gain instant insights into the performance and progress of their teams or departments.

For a director or a higher-level executive, time is often limited so the ability to comprehend complex information quickly becomes crucial. Static visualizations, most often in the form of a dashboard containing a small variety of metrics visualized, can enable them to cut through the noise and focus on the most important aspects of their organization. By presenting data in a visually appealing and condensed format, these visualizations eliminate the need for delving into extensive reports or spreadsheets, saving valuable time and effort.

The power of static visualizations lies in their ability to distill large amounts of data into meaningful and digestible chunks. A well-designed static dashboard snapshot can present key performance indicators, trends, and benchmarks in a visually compelling manner. For example, a director may use a bar chart to compare sales figures across different regions, instantly identifying the top-performing areas or spotting any significant disparities. A line chart can illustrate the revenue growth trajectory over time, giving a clear indication of the organization's overall progress.

Let's look at a simple example to *get the picture*. Here's a dashboard for our sales manager that gives them four main indicators of business performance:

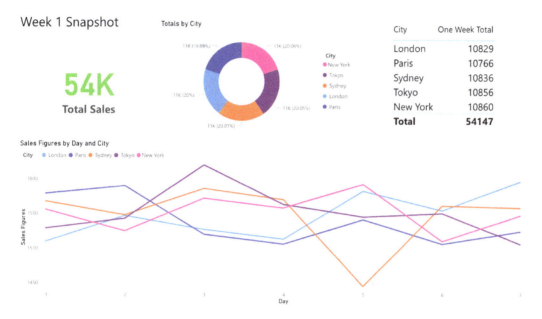

Figure 16.10 – Manager-level static dashboard visualization

From the top left, we have the following:

- The sales KPI. This tells the manager the totals for the stores represented. We'll say that these are the five flagship stores. A best practice with KPIs is to use color coding to indicate good or poor performance. The green indicates that the sales are on track with current projections.

- Next on the right is a simple donut chart, which indicates that all the stores are roughly equal in sales. Since this is a simple illustration, we're not going to get into whether this is good or bad, just that it's a super-quick visual representation and the manager will use it to infer and make decisions as needed.

- Then we have a simple table. For most use cases, it's a good idea to add a simple table where you can. Being able to look directly at numbers can provide an edge that some people still want to have, and with the limited amount of figures involved, it works. Don't use one where there's a very long list of figures, especially if there's not enough space to include it so that there's no scrolling need to see all the numbers. Up to 10 rows would be a good maximum to stick with because that amount is easy to scan quickly.

- Then, at the bottom, is our line chart, showing a comparison of the seven days of sales for each store, indicated by the five different colored lines. Here's where the power comes into play. This line chart very clearly shows an anomaly in Sydney. Notice the sharp dip, similar to what we saw in the line chart illustration from an earlier section, and because it varies from the trends of the other stores, this will allow the manager to alert an analyst to do further investigation by way of an interactive dashboard. *Remember this because we'll keep this storyline going when we show that.*

As you can see, static visualizations in the form of a collection of indicators and chart styles can provide a snapshot of the current state of affairs, making them ideal for periodic reviews, board meetings, or presentations to stakeholders. Directors can rely on these visualizations to quickly assess the health of their organization, identify areas of concern – such as that dip – or spot emerging opportunities. They can effortlessly absorb the information presented and make informed decisions based on the insights gained from the static visualizations.

Moreover, static visualizations enhance communication within the organization by providing a common visual language that all stakeholders can understand. Directors can easily share these visualizations with their teams, fostering a shared understanding of goals, performance, and priorities. Static visualizations are a powerful tool for aligning the entire organization, ensuring that everyone is on the same page and working toward a common objective.

However, it's also important to note that static visualizations have their limitations. They provide a static view of data, meaning that any changes or updates would require generating new visualizations. As a result, they're not suitable for real-time monitoring or situations where up-to-the-minute information is necessary. If these are prepared on a daily basis for managers, that's a great way to use them.

These visualizations condense complex data into visually appealing and easily interpretable representations, enabling directors to make swift and informed decisions. By leveraging the power of static visualizations, managers and directors can stay informed, drive organizational success, and effectively communicate with their teams and stakeholders.

Interactive visualizations

Often the foundation of static visualizations, interactive visualizations allow users to filter, click, and drill down into metrics, and manipulate data to create static versions. If you and your team have any type of business intelligence software, it will provide interactivity. Tools such as Power BI or Tableau are common examples, although there are many more.

Interactivity adds an extra dimension not possible in static visualizations and it's for that reason that they're usually designed, built, and delivered as applications. They offer users the ability to explore data, manipulate variables, and uncover insights on their own. With interactive elements such as clicking for drill-downs, filtering to desired results, sorting on variables, and hovering to uncover tooltips, users can dig deeper into the data and gain a more comprehensive understanding of complex relationships and patterns.

With this functionality not being possible with static visualization snapshots, interactivity supports exploratory analysis. It enables users to ask and answer specific questions by dynamically adjusting the view or parameters. Interactive visualizations are particularly valuable in scenarios where the data has multiple dimensions, requiring users to examine various aspects and correlations. They promote a more iterative and data-driven approach, allowing users to derive insights that may not be immediately apparent in static representations.

In order to show you this powerful way to use visualizations to deliver the fruits of your UX analytics labor, we're going to continue to rely heavily on screenshots from a test bed created for this book. Here's the front page of the dashboard used to deliver the interactivity:

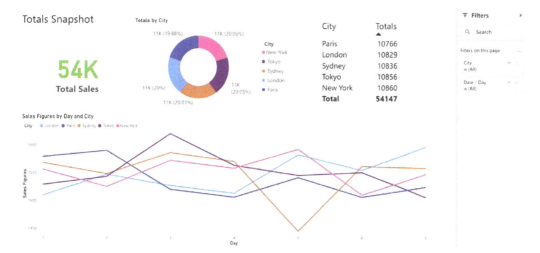

Figure 16.11 – Front page of the interactive visualization – dashboard

Now, you'll notice the inclusion of a filters panel to the right, and this is a page-level filter, meaning that when used, any visualization on the page that shares the attributes will be affected. Because we want to tell a very specific story with this dashboard, that is indeed the case. Also notice the use of labels on the sections and the use of legends to ensure that users understand what they're viewing quickly.

We have the KPI once more, the donut chart, the short tabular list, and the line graph. And now that we're in the application and have interactivity, we have the ability to hover for tooltips and to filter with the panel or by clicking on elements in the page.

Here's what we mean by *hover for tooltips*:

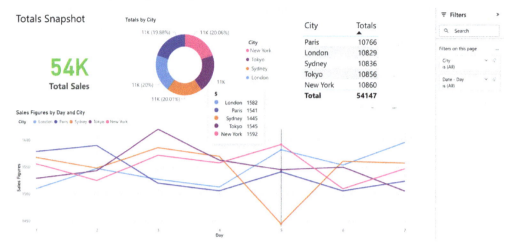

Figure 16.12 – Illustrating hovering for tooltips

This opened when I hovered the cursor over day five in the line chart. Nothing else happens to any of the other graphs because it's not a click event, just a mouse-over. What you see are the exact numbers for that day. If I move left or right, the popup changes to reveal the other days interacted with. These tooltips are also on the donut chart:

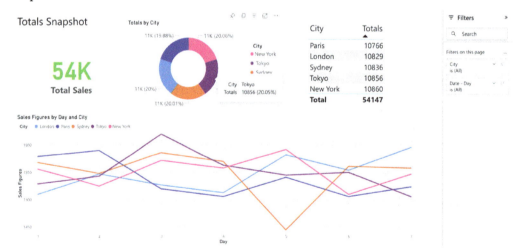

Figure 16.13 – Tooltips on the donut chart

Notice that the information is a bit different. If there were more variables available to us, we could include them here, but for now, we're keeping it simple. You get the name of the city, the total, and the percentage of the whole the total represents.

Now, let's click on Tokyo:

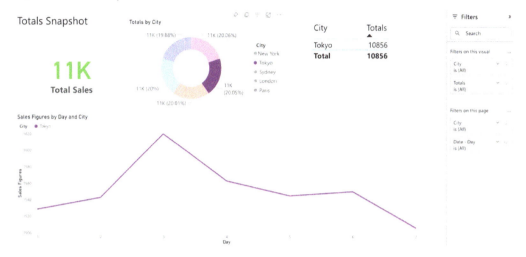

Figure 16.14 – Filtering click on Tokyo to isolate

I isolate the metrics on the page, in every chart and graph, to just Tokyo. Seems like there's a bit of a difference on day 3 that I never noticed before. I'd better check in with the store manager to see what's up. It's a good thing because the numbers went up so if they have a hot seller, maybe we should run it in the other stores. I also better make sure we have enough inventory.

Now let's see what we can do with the filters panel. Filters can normally be set to single select or multiple, and if we want to have the power to compare two stores, we need multiple select, which is the case in our demonstration.

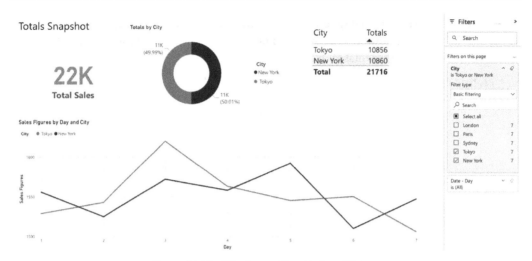

Figure 16.15 – Filtering on New York and Tokyo

Now we have a point of comparison. In our world, we'll say that New York and Tokyo are our two top stores, and even though on different sides of the world, they tend to run very evenly in terms of weekly sales and annual net revenue. We see here that for this weekly dataset, that holds true. If those two parts of the donut chart were unbalanced, we'd also see that on the line chart because the weekly figures would be much different.

Understanding levels of detail

Summary level one would be our use of the KPI but without the color indicator – just one big number that tells the user exactly what's happening in their quantitative world. And through your research and analysis work, KPIs are values you've discovered are important in your interviews. Of course, it's also possible that you work for a company that's been doing light UX and business intelligence work and therefore already have a list of KPIs that they watch.

Summary level two would add a color indicator to the KPI. We used green to let the manager know that it's on track. Yellow could mean that it's starting to trend toward trouble, and red would indicate trouble. It's always important to know whether up is good or up is bad. If, for instance, your metrics are counting the number of customer service ratings below a certain threshold, then up would be bad, and yellow or red. You must understand the context first. That makes sense, right?

Summary level two is also a level of comparison and is therefore an action-level summary. Comparison motivates us when incentives are in place that measure growth and/or effectiveness. In other words, are we successful or not? Comparison can define context, so it's important for you and your UX analytics team to understand what's in the minds of your users. Here are some typical questions when a user looks at your interactive dashboard:

- How does this visual match my responsibilities?

- Are there any strong trends here?

- Do I need to do anything at this moment, or can I wait?

- Am I on track to get a bonus this quarter?

- Why is our XYZ route going down like that?

Analytics charts that compare one value to another will give our users answers to those questions. Let's look at an example that gives a user context in sales figures (**YTD = Year to Date**, **LY = Last Year**):

Figure 16.16 – Summary level 2 comparison illustration

Notice this time that instead of a color indicator, I've introduced the icon to the right of our comparison-to-last-year figure. These certainly can be combined with color, but also consider *508 compliance* and the fact that this will let anyone know that the numbers are up slightly. If this number isn't on track with projections for this time of year, you could add a *minus* symbol underneath the up icon. And remember too that you can show these symbols in a legend for easy reference to anyone new viewing your charts.

As a summary level two visual, the user can now see where they are for the year. Sales projections are used by every company, and this now uses "where we are today" compared to "where we were last year" and the arrow indicates that we're over it, but not by much. Any percentage is of course subjective, so if your company does business in the hundreds of millions of whatever currency you use, 2.2% could be a lot of money.

Summary level two can also be used by way of 100% stacked bar charts. When used with the indicator line, as in this example, your users know where the breakeven point is, in conjunction with the visual indicator of one section being shorter or longer than the other. The colors here indicate planned revenue versus current revenue. If the blue is longer than the red, you're behind the plan and will need to do further investigation to figure out why and do what's needed to correct the situation.

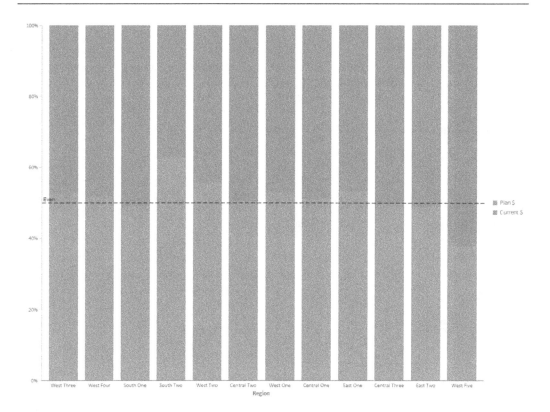

Figure 16.17 – Summary level 2, analysis level 1

Realize that this could just as easily be based on a count of positive versus negative reviews, the number of women versus men who completed a form, the number of service calls your reps performed in the East versus the West, and so on. The metric types won't matter – as long as you need to compare, and can count, this will work for you.

So, what's that about analysis level 1? That simple threshold that we included is an entry-level analysis technique because it gives an immediate visual cue for an analyst, manager, or whomever, to understand performance immediately. They can then make a report that Region South 2 is behind, and Region West 5 is ahead. Yes, it's still a summary, but because of the fact that we've set a threshold for performance, it now also gives us an analytics capability.

In *Analysis level 1*, we're going to start with the first level explanatory. This illustration shows a single metric across different groups – in this case, regions. Again, we see the threshold line as an indicator, but now you'll also notice something different – the colors are coded to what's over or under the line. We've used green to indicate that all is well with a given region, and pale orange to show that these regions are under-performing.

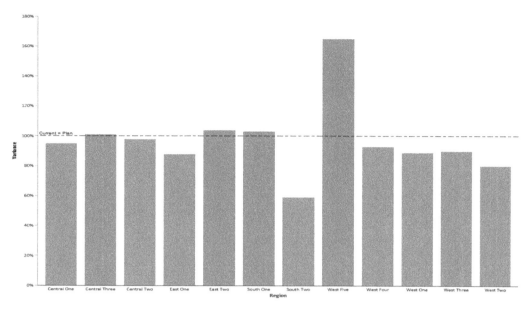

Figure 16.18 – Analysis level 1: first level explanatory example

In *Analysis level 2* (*Figure 16.19*), known as the second level explanatory, we dive deeper into the data by incorporating a more extensive set of metrics. In this case, we're using a scatter plot to compare three specific metrics, introducing a higher level of complexity and providing a clearer visual depiction of the differences. The metrics being compared are last year's data, the current year's data, and the planned data, all within the context of our sales analysis theme.

By examining the spread between the bubbles or data points on the scatter plot, we can gain valuable insights into the range or variance of each metric. This information becomes particularly valuable in sales analysis, as it helps identify areas where performance may be consistent, concentrated, or widely dispersed. The visual representation of the spread between the bubbles enhances our understanding of the data and facilitates the identification of patterns or anomalies that may not be immediately apparent in other types of visualizations.

Looking at the visualization, where there are regions with a tighter spread, we know things aren't terrible. Where you see larger spreads, such as in West Four, South One, and especially in South Two, you know immediately that something is out of whack and further analysis needs to be performed.

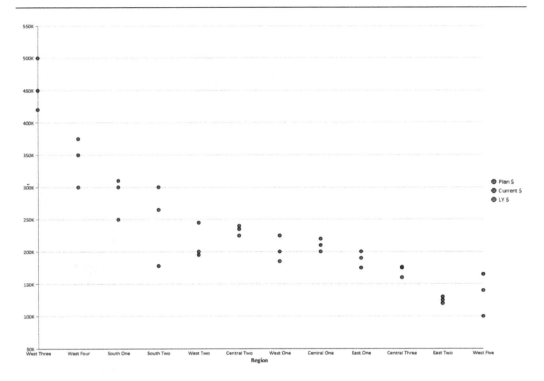

Figure 16.19 – Scatter plot analysis level 2

> **Note**
>
> I'll point out here that you should realize by now that these types of visualizations can help you create better reports for your users. This isn't only something you want to provide to them – you should be using data visualization to ensure that you're creating the best output possible.

What else can you quickly infer at this level of analysis? Think about my example of the large spreads, and figure out what else is portrayed in this scatter plot chart.

Throughout this section, we've explored four main types of data visualizations: bar charts, donut charts, line charts, and scatter plots. Each visualization type offers unique advantages and is suited for different analytical purposes. Let's summarize these types and their key characteristics:

- **Bar charts**: Bar charts are versatile visualizations that display categorical data using rectangular bars of varying lengths. They're ideal for comparing different categories or showing the distribution of a single variable. Bar charts provide a clear visual representation of data, allowing for easy identification of trends, comparisons, and outliers.

- **Donut charts**: Donut charts are effective for displaying the composition or proportion of different categories, enabling your users to understand the relative contribution of each category at a glance.

- **Line charts**: Line charts depict data using a series of points connected by lines and are particularly useful for showing trends and changes over time. They provide a visual representation of the relationship between variables, allowing for the identification of patterns, fluctuations, and correlations.

- **Scatter plots**: Scatter plots use individual data points plotted on a graph to show the relationship between two or more variables. They're valuable for identifying correlations, clusters, or outliers within a dataset, helping to identify patterns, trends, or anomalies that may exist.

By studying and mastering the characteristics and applications of these four main types of data visualizations, you and your team can select the most appropriate visualization method to present data effectively. Whether it's comparing categories, showing proportions, analyzing trends over time, or exploring relationships, these visualizations offer valuable insights and aid in making data-driven decisions.

Moving toward predictive and prescriptive analysis

While data visualization offers invaluable insights into historical data, it also has the power to pave the way for proactive business practices. In this final section, we'll explore how data visualization acts as a catalyst for predictive and prescriptive analysis. By leveraging advanced analytics techniques and harnessing the capabilities of machine learning in certain cases and contexts, we can move beyond reactive decision-making and embrace a proactive approach. We will delve into the transformative potential of data visualization in enabling us to anticipate future trends, simulate scenarios, and make informed decisions that shape the trajectory of our businesses.

Predictive analysis

Predictive analysis is the proactive process of using historical data, statistical algorithms, and machine learning to forecast future outcomes. If you've got a data science team to work with, this is the start of their involvement in your analytics practice and a strong path forward toward the digital acceleration of all your UX business practices.

Through the application of predictive analysis, organizations can tap into their vast data repositories to uncover hidden patterns, trends, and correlations that might not be immediately evident through traditional analysis methods. By integrating statistical algorithms and machine learning models, they can identify valuable insights and develop predictive models capable of anticipating customer behavior, market trends, and business performance.

The key here lies in the data repositories already at hand. A younger organization may not have sufficient data to start moving into predictive, so keep that in mind. In most cases – and there will always be exceptions – this will be easiest to accomplish in older organizations that have kept data and managed it well.

With a proficient data science team at their disposal, organizations can enhance their understanding of customer preferences, optimize resource allocation, and make well-informed decisions to drive business growth. The team's expertise in analyzing complex datasets and constructing accurate predictive models empowers organizations to anticipate customer needs, personalize user experiences, and optimize their product offerings.

This is not to say that you have to have a data science team. While they tend to make things easier because of their expertise, this can be done by you and your team. You'll need access to all the data at hand, so having a friend who's the data steward is helpful, as they can point you in the right direction, give you the correct permissions, and help you when things don't make sense.

Moreover, predictive analysis acts as a catalyst for digital acceleration within the realm of UX business practices. By harnessing the power of predictive models, organizations can proactively adapt to evolving market dynamics, identify emerging trends, and maintain a competitive edge. These insights enable businesses to make strategic decisions that maximize user satisfaction, increase customer engagement, and drive revenue growth.

Back to our data science team discussion again, the involvement of that team in predictive analysis not only enhances an organization's analytical capabilities but also fosters a culture of data-driven decision-making. Collaborating with the UX team, the data science team synergizes user-centric design principles with advanced analytics techniques. This collaboration allows the organization to leverage data-driven insights to refine the user experience, optimize product features, and create innovative solutions that meet the evolving needs of its target audience.

Data visualization plays a crucial role in this process by transforming complex patterns and trends into visually intuitive representations. With predictive analysis, organizations can identify potential opportunities, risks, and patterns that can significantly impact their operations. By visualizing predictions and projections, decision-makers gain a clearer understanding of the potential outcomes and can take preemptive actions to maximize opportunities or mitigate risks.

When you don't have data scientists at your disposal

If there's no team, you can do it on your own. Because you don't have that team, that will generally mean that you're in a company that's either very old and still kind of old-school, or so new that the fact that they have you and your UX team doing analytics to help move them forward is refreshing. It means they don't yet understand the power of the data at hand and what predictive analytics can do for them.

Let's work from the premise that you have years of data at your disposal. With this in mind, even if it's not your reality, we can create a plan of action to move toward predictive. What does predictive mean? It means that by way of the right type of analysis and data usage, you can *predict* the likelihood of an occurrence of some sort within your business. I can almost guarantee you that if you're in a large company that's well established in your market, your sales force is already predicting certain things.

Marketing teams, when they study trends and research deep into historical data, are great for doing predictive analysis. In your practice as a UX professional, especially at today's 400 miles-a-minute pace, your ability to support your management teams or consult customers with information that will allow them to make decisions before it becomes reactive, is monumental. In fact, I'll go as far as to say that if you aren't starting to push this agenda now, with everything else you've learned in this book, you'll get left in the dust by your competitors who know this is the beginning of the future of business.

This is such a huge topic that it will be impossible to cover it all in the remainder of this book, but we're going to get you started on the fundamentals so that you can get started and show your users what a powerful practice this is to start right away. Here's an ordered list of the steps with concise notes for each step:

1. **Assess your company's situation**: If you don't have data scientists available, it suggests that your company is either old and slow to catch up with modern business practices, or relatively new and unfamiliar with the potential of data and predictive analytics.

2. **Plan for the future**: Even if you currently lack the necessary resources, you can develop a plan to move toward predictive analytics. The goal is to leverage analysis and data usage to predict likely occurrences within your business.

3. **Leverage existing data**: Assuming you have access to years of data, focus on utilizing this valuable resource. This data can be used to make predictions about various aspects of your business, such as sales forecasts, with the potential to improve decision-making. This is the most important factor, so we'll follow this list with a section explaining.

4. **Utilize marketing teams**: Marketing teams, especially those well versed in studying trends and analyzing historical data, are excellent resources for conducting predictive analysis. Collaborating with them can enhance your ability as a UX professional to provide valuable information for proactive decision-making.

5. **Embrace a proactive approach**: In today's fast-paced business environment, it's crucial to support management teams or consulting clients with predictive insights. This enables them to make informed decisions before they become reactive, giving you a competitive advantage.

6. **Recognize the significance of predictive analytics**: Understanding the importance of predictive analytics is essential for your professional growth. By embracing this practice and staying ahead of the curve, you position yourself and your organization for future success.

7. **Focus on the fundamentals**: While the topic of predictive analytics is vast, it's important to start with the basics.

Leveraging existing data

When it comes to leveraging and utilizing existing data for predictive analytics, it's crucial to follow a systematic approach to ensure accurate predictions and gain confidence in the insights you and your team provide to your customers. Here are the key steps to guide you and your team in this process:

1. **Identify relevant data sources**: Start by identifying the data sources within your organization that are most relevant to your predictive analytics goals. This may include customer data, sales records, marketing campaigns, website analytics, or any other data that can provide insights into the variables you want to predict. Find the person who owns or manages the data and work with them to ensure you're getting the right data.

2. **Cleanse and preprocess the data**: Before you can extract meaningful insights, it's important to clean and preprocess the data. This involves handling missing values, dealing with outliers, normalizing data, and ensuring consistency across different datasets. By cleaning the data, you enhance its quality and reliability.

3. **Define the prediction target**: Clearly define what you want to predict. This could be customer churn, sales revenue, product demand, or any other measurable outcome that aligns with your business objectives. Defining the prediction target helps focus the analysis and ensures the right variables are considered:

    I. Go back to your stakeholders first and discuss it with them. If necessary, add an addendum to your North Star statement, or pitch this as a new project with the intent of deepening the understanding of historical data so that everyone can be more efficient.

    II. Get everyone on the team comfortable with this new process and collaborate across as many teams as you can, include product and engineering when possible.

4. **Identify relevant predictors**: Explore the available data to identify potential predictors or independent variables that might have a significant impact on the prediction target. This requires domain knowledge and an understanding of the factors that influence the outcome you're trying to predict. Identify KPIs up front and vet them with the stakeholder. If you're starting to see data that's been buried and forgotten, document and communicate that because resurrecting older data can sometimes be the deciding factor in the success of predictive analytics.

5. **Analyze and visualize the data**: Apply exploratory data analysis techniques to gain insights and identify trends within the data. Utilize data visualization tools to present the patterns and relationships discovered. Visualizations can help stakeholders understand the data better and make informed decisions.

6. **Build predictive models**: Select appropriate machine learning algorithms based on the nature of the prediction problem and the available data. Train these models using historical data and evaluate their performance using suitable metrics such as accuracy, precision, recall, or mean squared error. Iterate and fine-tune the models to improve their predictive capabilities. This will of course be dependent on a data science team or someone who knows how to work with machine learning.

7. **Validate and test the models**: Use separate test datasets to validate the predictive models. This helps assess their performance on unseen data and ensures that the models generalize well. Validate the models using techniques such as cross-validation or holdout validation to avoid overfitting.

8. **Make predictions and monitor results**: Once the models are validated, deploy them to make predictions on new data. Monitor the results regularly and compare them against the actual outcomes. This allows you to assess the accuracy and reliability of the predictions and refine the models if necessary.

If you can't do machine learning, you can use business intelligence tools to get where you need to go. Because you now have a much larger pool of data to work with, you and your stakeholder can determine what to go after, and by ingesting all the data into your tool, you can create interactive dashboards to filter, rework, and refine until you've been able to show trends strong enough to start making predictions.

Some hints: Look for trends during events that don't happen every year, such as political elections, the World Cup, the Olympics, global economic summits such as G7 or G20, natural disasters, and space missions. When you have enough historical data, these events will show in your analysis as patterns that are different than those from normal business numbers.

By following the steps listed and leveraging existing data effectively, you and your teams can start spotting trends and making predictions with confidence. When you can, seek assistance from data scientists or analytics experts who can provide guidance throughout the process and help to extract maximum value from the available data.

Remember, the power of predictive analytics lies in the ability to uncover valuable insights from data, anticipate future outcomes, and make informed decisions that drive business success.

Prescriptive analysis

Prescriptive analysis takes proactive decision-making a step further by not only predicting future outcomes but also prescribing the best course of action to achieve desired results. By combining predictive models with optimization techniques, prescriptive analysis enables organizations to simulate various scenarios and evaluate the potential impact of different decisions.

> **Note**
> So, predictive is "this is very likely to happen," and prescriptive is "this is very likely to happen, and this is what we should do about it now."

Data visualization is instrumental in presenting these simulated scenarios, allowing decision-makers to compare and contrast different strategies visually. Through interactive visualizations, decision-makers can assess the potential outcomes of each scenario, identify the most optimal course of action, and make data-driven decisions that drive their businesses forward.

For companies that lack data scientists or the infrastructure to implement machine learning, there are still valuable avenues to explore in harnessing the power of prescriptive analysis. By combining business intelligence tools and design thinking principles, organizations can unlock insights and drive proactive decision-making to achieve desired outcomes.

Business intelligence tools provide a user-friendly interface and a range of analytical capabilities that allow non-technical users to explore data, uncover patterns, and gain valuable insights. These tools often offer pre-built analytical models, customizable dashboards, and interactive visualizations that enable stakeholders to make data-driven decisions without the need for complex algorithms or coding.

To effectively utilize business intelligence tools for prescriptive analysis, organizations can adopt design thinking principles. Design thinking focuses on understanding user needs, ideating creative solutions, and iteratively refining them based on feedback. By applying design thinking to the analysis process, companies can involve stakeholders and subject matter experts, collaborate on defining objectives and constraints, and iteratively refine recommendations to ensure they align with business goals and stakeholder expectations.

Here are key steps to effectively leverage business intelligence tools and design thinking for prescriptive analysis:

1. **Identify business objectives**: Clearly define the desired outcomes and **key performance indicators (KPIs)** that align with the organization's strategic goals. This provides a focus for the analysis and ensures that the recommendations derived through prescriptive analysis contribute to tangible business improvements.

2. **Engage stakeholders**: Involve stakeholders from different departments and levels of the organization in the analysis process. Conduct workshops, interviews, or surveys to gather insights into their needs, pain points, and aspirations. This collaborative approach ensures that the recommendations generated through prescriptive analysis are relevant and address real-world challenges.

3. **Gather and prepare relevant data**: Identify and gather the necessary data sources that provide insights into the factors influencing the desired outcomes. Be sure to cleanse and preprocess the data, ensuring its accuracy and consistency so that it's trustable. Business intelligence tools often have data integration and preparation capabilities that simplify this process for non-technical users.

4. **Visualize and explore the data**: Utilize the visualization capabilities of the business intelligence tools to present the data in a meaningful and interactive way. Encourage stakeholders to explore the data, ask questions, and identify patterns or trends that may inform decision-making. This collaborative exploration helps generate valuable insights and sparks discussions among stakeholders.

5. **Ideate and refine recommendations**: Facilitate ideation sessions with stakeholders to brainstorm potential actions or strategies based on the insights gained from the data analysis. Encourage diverse perspectives and foster an environment of creativity. Iteratively refine these recommendations based on feedback, feasibility, and alignment with business objectives.

6. **Communicate findings effectively**: Use compelling visualizations and storytelling techniques to communicate the insights and recommendations derived from the prescriptive analysis. Tailor the communication to the needs and preferences of different stakeholders, ensuring clarity and relevance. Highlight the potential impact of recommended actions and articulate the value proposition for stakeholders.

By following these steps and leveraging business intelligence tools alongside design thinking, companies without data scientists or machine learning capabilities can unlock the power of prescriptive analysis. This approach empowers stakeholders to make data-informed decisions, optimize processes, and drive positive business outcomes, ultimately enabling organizations to thrive in today's data-driven business landscape.

In a nutshell, prescriptive analysis helps organizations move from reactive decision-making to proactive decision-making. By leveraging historical data and applying advanced analytics techniques, prescriptive models provide decision-makers with actionable recommendations to shape future outcomes. These recommendations not only optimize current processes and strategies but also enable organizations to anticipate and adapt to changing market dynamics. By embracing prescriptive analysis, businesses can gain a competitive edge by making data-driven decisions that lead to better resource allocation, improved customer satisfaction, and sustainable growth.

Summary

Data visualization empowers decision-makers to navigate the complexities of business today by aligning with the visual nature of our brains. In the ultimate goal of understanding and utilizing predictive and prescriptive analysis, data visualization becomes your champion by providing intuitive and actionable insights to work from.

Through visually compelling dashboards, decision-makers can access real-time updates on key performance indicators, predictive models, and scenario simulations. Interactive visualizations allow them to explore different variables, drill down into details, and gain a deeper understanding of the factors influencing future outcomes. By interacting with the data visually, decision-makers can identify trends, patterns, and relationships that may not be apparent through traditional numerical analysis alone.

The adoption of predictive and prescriptive analytics, supported by data visualization, is reshaping business practices across industries, on every continent of the globe. It allows organizations to become proactive rather than reactive, enabling them to anticipate market shifts, optimize operations, and seize emerging opportunities. By visualizing data-driven insights, decision-makers can communicate complex information across the organization, align stakeholders, and drive a culture of data-driven decision-making.

It's nearly impossible to stress how important the lessons in this chapter are for you to learn and master. By using UX analytics as prescribed, you allow stakeholders to assess the situation with confidence. This helps them make the best decisions when it comes to costs and revenue in the context of business goals. This, combined with your operational analytics, creates a solid foundation for your business to grow.

In our final chapter coming up next, we'll talk about heuristics, and that will tie all of this learning together for you.

17
Heuristics – How We Measure Application Usability

While heuristics refers to mental tricks we employ to find shortcuts to common tasks so that we reduce the cognitive load on our brains, heuristic evaluations seek to assess an application's mirroring of human responses to work. In the context of usability evaluation, they're a set of principles or guidelines that serve as a framework for assessing the usability of a product.

These guidelines are derived from extensive research and best practices in the field of user experience design, and they provide a quick and effective way to identify potential usability issues and areas of improvement. By utilizing heuristics evaluations, you can gain valuable insights into how well your application aligns with human cognitive processes and create a more intuitive and user-friendly experience. Consider this work to be fundamental to an intelligent and proactive UX analytics practice.

This subject area was left for last because it not only ties together the diverse facets of UX analytics but also provides a comprehensive framework for enhancing user experiences, the problems of which were uncovered by your analytics work. This arrangement ensures that you and your team can leverage the full breadth of insights gained from the previous chapters to optimize your approach to UX analytics and ultimately create exceptional user-centric products and services.

With all that in mind, this chapter will lay out the fundamentals, get you involved with one of the most well-respected think tanks on user experience and usability, and explain why this final chapter ties everything else from this book together. Here are the sections:

- Getting to know application heuristics
- Exploring the Nielsen Norman Usability Heuristics Platform
- Navigating through a few types of evaluation

Getting to know application heuristics

Heuristic evaluation is an invaluable discipline that, if done correctly, helps identify usability problems early in the design process, resulting in a more user-friendly and efficient product. Whether done during initial design or employed to help explain your analytics findings and fix usability problems, you can objectively assess the quality of the user interface and identify any pieces or workflows that may weaken user satisfaction, efficiency, and learnability.

In the design phase, heuristic evaluation serves as a potent tool to detect potential usability issues before the product is released to users. Designers can systematically analyze the interface and interactions using a set of predefined protocols or steps, ensuring adherence to established usability principles. This proactive approach not only saves valuable time and resources but also mitigates user frustration and dissatisfaction that may arise from poor usability.

One of the key advantages of heuristic evaluation is its ability to uncover both obvious and subtle usability problems. While some issues may be glaringly apparent, such as confusing navigation or unclear error messages, heuristic evaluation also brings attention to more nuanced usability concerns. These could include inconsistencies in visual design, non-standard use of controls, or excessive cognitive load imposed on users. By identifying and addressing these subtler issues, the overall user experience can be significantly enhanced.

Let's look at a few examples. I'm going to teach this in terms of "interaction cues" and "design metaphors." An interaction cue is a visual hint that allows a user to know what to do next. A design metaphor is a visual indicator that sparks instant recognition in the user. If each is heuristically sound, the user understands what's required, with no need for secondary thought or determination; the path is obvious. These few examples will show instances where a heuristics evaluation would flag the need for correction.

Remember in the data visualization chapter where we added the up arrow with color to the KPI? That's a design metaphor because it combines the up arrow, which signifies a rise, with the color that signifies it's a good thing. Remember too that that is contextual.

What about this?

Figure 17.1 – Broken heuristics example 1

This would completely trip up a user because their expectation is that green always means go and red always means stop. A heuristics evaluation would catch this and flag it for fixing, quite obviously. In fact, the mind of most users would, without even looking at the text on the button, cause them to click the green button if they wanted to continue. Therefore, the interaction cue – and, one could argue, the design metaphor – is broken.

Think about this: when you put the salt back in the cupboard, you put it in roughly the same spot each time. If someone comes in behind you and moves it, chances are that you won't find it when you go back to look for it. Your brain – visual brain, remember? – places it in a specific spot and expects it to be there each time.

Here's another one:

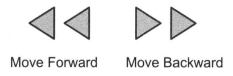

Move Forward   Move Backward

Figure 17.2 – Broken heuristics example 2

In many cultures around the world, this would be confusing because pointing to the right means to move forward and to the left means to move backward. Though kind of a silly example, we paired text with affordances, but the text is a label outside the image, and this makes it a different use case of heuristics. Again, we have a broken interaction cue. Let's take this one step further. What if the action of using the arrows that are labeled **Move Backward** actually took you forward? Big misdirection, right?

Alright, let's move on. Heuristic evaluation provides a structured framework for assessing usability, enabling evaluators to methodically examine different aspects of the user interface. This approach promotes consistency and ensures comprehensive coverage of usability considerations. By evaluating the product against established heuristics, evaluators can not only identify existing usability problems but also provide actionable recommendations for improvement. This iterative refinement process leads to an optimized user experience.

To reiterate a point, heuristics evaluations can be done at any point in time, on any application. The ideal time is in the iterative phases of design and development, but heuristics can certainly be carried out after a product is complete or when UX analytics work starts to prove that user satisfaction is low and improvements need to be made to bring that satisfaction rating back up.

Exploring the Nielsen Norman Usability Heuristics Platform

When it comes to heuristics, one of the most renowned frameworks is the Nielsen Norman Group's 10 usability heuristics. Developed by usability experts Jakob Nielsen and Rolf Molich, these heuristics have become a staple in the field of user experience evaluation. Let's take a closer look at each of these heuristics.

Visibility of system status

Users should be provided with clear and timely feedback about the system's response to their actions so that they feel in control and able to make the best decisions when needed. Here are four of the most common:

- **Progress indicators**: These show users how far along they are in a task, and they also help to keep users informed about how long a task is likely to take

- **Error messages**: These should be clear and informative, and they should tell users what went wrong and how to correct the error

- **Notifications**: These can be used to inform users about important events, such as when a file has been uploaded or when a new message has been received

- **Visual cues**: These can be used to indicate the status of the system, such as a green checkmark to indicate that a task has been completed or a red X to indicate that an error has occurred

Match between the system and the real world

The system should speak the users' language, using terms, concepts, and metaphors familiar to them. This means that the system should use language that's easy to understand and that's consistent with the users' expectations. Here are some examples:

- **Use familiar terminology**: This means avoiding jargon or technical terms that users may not understand. For example, a website that sells clothes should use terms such as "shirt" and "pants" instead of "garment" and "apparel."

- **Use real-world metaphors**: The system can use real-world metaphors to help users understand how it works. For example, a file manager could use a metaphor of a folder to represent a collection of files. This can make the system more intuitive and easier to use.

- **Follow real-world conventions**: This means using the same conventions for things such as button placement, menus, and error messages. This can help users to learn the system more quickly and make it easier to use.

- **Use recognizable icons**: Using recognizable icons to represent actions or objects can make the system more visually appealing and easier to use. For example, a website that sells clothes could use icons of a shirt and a pair of pants to represent the "shirts" and "pants" categories.

User control and freedom

Users should have the ability to undo and recover from errors or unintended actions. Giving users this ability can help to reduce frustration and improve the user experience. It can also help to prevent users from making irreversible mistakes. Let's look at these examples:

- Provide clear and visible exit points. Users should be able to easily find a way to exit a task or process, even if they've made a mistake. This could be a "**Cancel**," "**Back**," or "**Close**" button.

- Allow users to undo their actions. Users should be able to undo their actions if they make a mistake. This is especially important for irreversible actions, such as deleting a file or submitting a form.

- Support multiple levels of undo. Users should be able to undo multiple actions, so they can go back in time and correct their mistakes.

- Allow users to customize their experience. Users should be able to customize their experience, such as by choosing the default settings or by changing the look and feel of the interface.

- Provide clear and concise error messages. If an error occurs, users should be given a clear and concise error message that explains what went wrong. This will help users understand the error and take steps to correct it.

Consistency and standards

The system's design should follow established conventions and brand standards to ensure a familiar and predictable user experience. This means that the system should use the same terminology, icons, and actions throughout the system. This will help users learn the system more quickly, and it will make the system more predictable and easier to use. Here are five examples:

- **Using the same terminology throughout the system**: This means using the same words and phrases to refer to the same things. For example, if a website uses the term "account" to refer to a user's profile, it should use the term "account" throughout the website.

- **Using the same visual elements throughout the system**: This means using the same colors, fonts, and icons throughout the system. This helps users to recognize and understand the different elements of the system.

- **Following the same conventions throughout the system:** This means using the same conventions for things such as button placement, menus, and error messages. This helps users to learn the system more quickly and make it easier to use.

- **Grouping similar elements together**: This helps users to find the information or features they are looking for. For example, a website could group all of its product listings together in one section.

- **Using clear and concise labels**: This helps users to understand what each element of the system does. For example, a button should have a label that clearly explains what will happen when the button is clicked.

Error prevention

The system should be designed in a way that prevents errors or guides users in recovering from them. For example, if a user accidentally deletes a file, the system could provide the user with instructions on how to recover the file. Take a look at these examples:

- **Using good defaults**: Provide users with sensible defaults for things such as input fields and settings. This can help to prevent users from making mistakes. For example, a website could use a default password that is easy to remember but difficult to guess.

- **Providing helpful constraints**: Limit the range of values that users can enter into input fields. This can help to prevent users from entering invalid data. For example, an application could limit the number of characters that users can enter into a username field.

- **Validating input**: Check the validity of user input before it is processed. This can help to prevent errors from occurring. For example, a website could validate a user's email address before it's used to create an account.

- **Providing warnings**: Warn users about potential errors before they occur. This can help users to avoid making mistakes. For example, a website or application could warn users if they're about to delete a file that's still in use.

- **Providing undo/redo**: Allow users to undo their actions if they make a mistake. This can help users to correct errors without having to start over. For example, a word processor could allow users to undo their last action by clicking on an "undo" button.

Recognition rather than recall

Minimize the user's memory load by making objects, actions, and options visible and easily recognizable. For example, an application that uses icons to represent different actions will make it easier for users to identify the actions they want to take than an application that uses text labels for the actions. This is because users' visual brains can recognize icons more easily than they can remember text labels. Examples are as follows:

- **Using icons**: Icons are a great way to help users recognize information. For example, a website could use icons to represent different actions, such as "save," "delete," and "print."

- **Using labels**: Labels are also a great way to help users recognize information. For example, an application or website could use labels to identify different input fields, such as "*username*" and "*password*."

- **Using menus**: Menus can help users to recognize and find the information they are looking for. For example, a website could use a menu to organize its content into different categories, such as "*products*," "*services*," and "*contact us*."

- **Using search**: Search can help users to find information that they may not be able to recognize. For example, an application could provide a search bar that allows users to search for specific keywords.

- **Using feedback and confirmation**: Feedback and confirmation can help users to recognize the status of their actions. For example, an application could provide feedback when a user enters a valid username or password.

Flexibility and efficiency of use

Accommodate both novice and expert users, allowing them to accomplish tasks using different levels of expertise. For example, a system could allow novice users to complete tasks by following a step-by-step wizard, while allowing expert users to complete tasks by using shortcuts and keyboard commands. The system could also provide users with the ability to customize the interface to their own preferences. Take these examples, for instance:

- **Allow users to customize the system**: This gives them the flexibility to tailor the system to their own needs and preferences. For example, a word processor could allow users to change the font size, the color scheme, and the keyboard shortcuts.

- **Provide shortcuts**: This gives them the efficiency to quickly access frequently used features. For example, a web browser could provide shortcuts for opening new tabs, closing tabs, and navigating to different websites.

- **Allow users to undo and redo their actions**: This gives them the flexibility to experiment and make mistakes without fear of losing their work. For example, a spreadsheet program could allow users to undo their last action by clicking on an "undo" button.

- **Provide help and documentation**: This gives them the efficiency to learn how to use the system and troubleshoot problems. For example, a software program could provide a help menu that explains the different features of the program.

- **Allow users to work in different ways**: This gives them the flexibility to approach tasks in a way that suits their individual working styles. For example, a word processor could allow users to type text, dictate text, or use a handwriting recognition system.

Aesthetic and minimalist design

Strive for simplicity and clarity in the user interface, removing unnecessary elements and distractions. Aesthetic and minimalist design is important for usability because it can make systems more user-friendly and enjoyable to use. When systems are simple and clear, users are less likely to be confused or frustrated. This can lead to increased productivity and satisfaction. Here are real-world examples for you:

- **Apple's product design**: Apple's products are known for their clean, simple, and elegant design. The iPhone, for example, has a very minimalist design with few buttons or ports. This makes the phone easy to use and visually appealing.

- **Google's Material Design**: Google's Material Design is a set of design principles that emphasize simplicity, clarity, and efficiency. Material Design is used in many Google products, such as the Google Search app and the Android operating system.

- **The website of Airbnb**: Airbnb's website is a great example of simple, clean, minimalist design. The website uses simple typography, white space, and high-quality images to create an uncluttered look. This makes the website easy to navigate while being visually appealing.

- **The Calm app**: Calm is a meditation app that uses minimalist design to create a calming and relaxing experience. The app's interface is simple and easy to use, and the app uses soothing colors and images to create a peaceful atmosphere.

Help users recognize, diagnose, and recover from errors

Error messages should be expressed in plain language, precisely indicating the problem, and suggesting solutions. For example, an error message that says **Error: Invalid input** isn't helpful at all. It doesn't tell the user what's wrong with the input, or how to correct the error. A better error message would say something like **Error: The value you entered is not a valid number. Please correct by entering a valid number.** Here are some examples:

- **Use clear and concise error messages**: Error messages should be easy to understand and should clearly explain the problem that has occurred. See the previous example.

- **Provide context for error messages**: Error messages should be provided in the context of the user's current activity. For example, if a user is trying to log in and they receive an error message, the error message should be specific to the login process and tell them what the problem is.

- **Suggest solutions to errors**: Error messages should suggest solutions to the problems they're describing. For example, an error message that says "**The username you entered is not valid**" could suggest that the user try entering a different username.

- **Use visual cues to indicate errors**: Visual cues, such as red text or icons, can be used to indicate errors. This can help users quickly identify and diagnose errors.

- **Allow users to recover from errors**: Users should be able to recover from errors without having to start over. This can be done by providing users with the ability to undo their actions or by providing them with a way to correct the errors they've made.

Help and documentation

Provide relevant help and documentation, easily accessible and tailored to users' needs, to assist them in accomplishing their tasks. Help and documentation are important for usability because they can help users learn how to use the system and troubleshoot problems. When help and documentation are easy to find and use, users are more likely to be able to use the system effectively. This can lead to increased productivity and satisfaction. Here are a few examples:

- **Help menus**: Help menus are a common way to provide help and documentation for users. Help menus typically provide a list of topics that users can click on to get more information.

- **Online documentation**: Online documentation is a great way to provide help and documentation for users. Online documentation can be accessed from any device, and it can be updated easily.

- **Tutorials**: Tutorials are a great way to teach users how to use a system. Tutorials typically walk users through the steps of completing a task.

- **FAQs**: FAQs (frequently asked questions) are a great way to answer common questions that users may have. FAQs can be organized by topic, making it easy for users to find the information they need.

- **Video tutorials**: Video tutorials are a great way to provide visual instructions for users. Video tutorials can be especially helpful for users who learn best by watching.

You'll notice two points, 5 and 9, that talk about errors. At first, these two seem to be saying essentially the same thing because they are a bit ambiguous, so what's important in my opinion is to recognize that error prevention through the best design decisions should always be your first defense against user error.

So, consider number 5 to be about prevention through design, and number 9 to be about ensuring that users know how to get out of a jam, because no matter how much effort we put into good design, users are bound to make mistakes. Let's make sure we help them get out and back to productivity quickly.

Let's wrap all that up into a quick summary. At the platform level, Nielsen Norman's heuristics provide a set of guidelines for evaluating user interface designs. These guidelines include ensuring that the system's status is clearly visible to users, creating a match between the system and the real world, offering users control and freedom in their interactions, maintaining consistency and following established design and branding standards, preventing errors and providing clear error handling mechanisms, reducing users' cognitive load by presenting information in a visible and accessible manner, accommodating the experience levels of different users with shortcuts and customization, employing a clutter-free and minimalist design approach, providing clear error messages and recovery options, and offering helpful and context-sensitive documentation to assist users.

These heuristics serve as valuable principles for enhancing usability and creating user-friendly interfaces. By familiarizing yourself with these heuristics, you gain a powerful tool for evaluating usability issues within your product or system. Always keep in mind that when users are in your application or website, they're there because they need to accomplish something, and they bring expectations with them that you must meet, so use these 10 principles at the beginning if possible so that the design is as usable as possible.

While the Nielsen Norman Heuristics Platform provides a great starting point as well as a solid list of guidelines by which to design or evaluate a design for fixes or enhancements, there are other ways to do evaluations. Let's explore these next.

Making use of other types of evaluation

A key lesson to learn, no matter how or why you and your team may be asked to do evaluations, is to always think through the mind of your users. What might they expect? What might cause them to get lost or confused by the design?

The ultimate design is one that they have no immediate control over so you must do your best to help the design teams, the developers and engineers, and your business partners to understand and produce. Your UX analytics methods and practices all culminate right here. Either you're pulling from historical analytics to guide a design team right out of the gate for a new product, or you're using all the tools and processes written in this book to help fix problems brought to light by your work.

So, with that in mind, let's get into a few more types of evaluations that will round out your UX toolbox and make you and your team the group that people rely on to ensure that only the best products go to market.

Cognitive walk-through

This method involves analyzing a user's thought processes and decision-making as they use your application, and focuses on how well the system supports their goals and actions.

The cognitive walk-through is a powerful method used in UX design to understand how users think and make decisions while interacting with the application. It allows us to put ourselves in the shoes of the users and empathize with their thought processes, helping us uncover usability issues and improve the overall user experience.

During a cognitive walk-through, we carefully analyze the interface from the user's perspective, step by step, as they navigate through different tasks. We consider their goals, motivations, and expectations at each stage of interaction. By doing so, we gain valuable insights into whether the system effectively supports users in achieving their objectives and making informed choices.

As evaluators, we put ourselves in the mindset of a typical user – think persona – and ask critical questions along the way. We focus on the clarity of instructions, the visibility of available options, and the system's response to user actions. We pay close attention to the system's affordances – how well they guide users and provide interaction cues for the next steps.

By following this method, we can identify potential stumbling blocks or areas where users may struggle. We observe whether the interface provides the necessary information, feedback, and guidance for users to complete their tasks successfully. This allows us to detect any gaps in the system's support and make informed recommendations for improvement.

The cognitive walk-through combines empathy with expertise, as it requires a deep understanding of both the user's perspective and the principles of good design. By adopting this method, we gain valuable insights into users' cognitive processes and decision-making, helping us create interfaces that are intuitive, efficient, and enjoyable to use.

In summary, the cognitive walk-through allows us to step into the minds of our users, understand their thoughts and decision-making steps, and identify areas for improvement in the interface. It empowers us to make informed design decisions that enhance the user experience and create interfaces that align with users' mental models.

Feature inspection

Another valuable method used in UX design, feature inspection is used to carefully examine specific features or functions of an interface, ensuring they align with usability principles and guidelines. It allows us to dive deep into the details and intricacies of individual elements, empowering us to identify strengths, weaknesses, and opportunities for improvement.

During a feature inspection, we meticulously analyze each feature, component, or interaction within the interface. We put ourselves in the shoes of the users and empathize with their needs, expectations, and goals while interacting with these specific elements of our application. By doing so, we gain a comprehensive understanding of how well the features are designed and whether they effectively support user tasks and objectives.

As evaluators, we must understand the importance of designing features that meet users' needs while also adhering to established usability standards. We pay attention to the clarity of labels, the intuitiveness of controls, and the consistency of interaction patterns. We examine how well the features align with users' mental models, ensuring that they are intuitive and easy to use.

During this process, we may uncover areas where the interface excels, providing delightful and efficient user experiences. We will also identify areas that may require improvement, such as confusing or non-intuitive features that may hinder user productivity or satisfaction. Through this rigorous examination, we can make authoritative judgments based on our expertise and knowledge of best practices in UX design.

> **Note**
>
> Also extremely important is our dedication to reporting both the good and the bad so that we create a picture of the overall use of all features. Of course, the objective is to find the pain points and report those so that usability problems can be addressed and fixed, but it's also important for everyone involved to know where good design decisions were made, and where users are happy with the experience.

Feature inspection allows us to address specific design challenges and fine-tune individual components for optimal usability. By focusing on the details, we can enhance the overall user experience and ensure that every feature contributes to the interface's effectiveness and efficiency.

In summary, feature inspection involves a meticulous examination of specific features, evaluating their usability and alignment with user needs. Through this process, we can identify areas for improvement and make informed design decisions that result in an interface that's user-friendly, intuitive, and delightful to interact with.

Pluralistic walk-through

This approach involves a collaborative evaluation process, where a group of evaluators with diverse backgrounds and expertise collectively review the interface to identify usability issues. In my opinion, this method is the culmination of a truly collaborative UX analytics practice. Because it involves people with differing responsibilities in the design and production of an application, it accounts for a variety of mental models. It recognizes the value of different viewpoints and expertise, fostering a holistic understanding of the user experience and uncovering a wider range of usability issues.

During a pluralistic evaluation, a group of evaluators with varied backgrounds, skills, and knowledge come together to collectively review and assess the interface. Each evaluator brings their unique perspective, including designers, researchers, developers, and even end users. This diversity ensures a comprehensive examination of the interface from multiple angles, enriching the evaluation process with different viewpoints and expertise.

As evaluators, we approach pluralistic evaluation with empathy and respect for each participant's expertise – this is a safe space where we practice nonjudgement. We value the insights and perspectives that each member of the evaluation team brings to the table. Through open and constructive discussions, we encourage the sharing of observations, opinions, and recommendations, creating a collaborative atmosphere that fosters learning and growth.

Pluralistic evaluation empowers us to uncover a wide range of usability issues that may have gone unnoticed in an individual evaluation. The combined expertise of the evaluators helps identify strengths and weaknesses in different aspects of the interface, including visual design, interaction patterns and cues, and overall user flow. It allows us to surface diverse perspectives on user needs, preferences, and pain points, enabling a more comprehensive understanding of the user experience.

By engaging in pluralistic evaluation, we gain a deeper appreciation for the complexity of user interactions and the challenges our users may face. As a holistic group of professionals, we begin to recognize the importance of designing interfaces that cater to a diverse range of users, considering their varying abilities, backgrounds, and contexts of use.

In summary, pluralistic evaluation embraces collaboration and inclusivity, fostering a multidimensional approach to interface evaluation. By leveraging the collective expertise of a diverse group, we gain a broader understanding of the user experience and identify a wider array of usability issues. Through empathetic and constructive discussions, we collectively work toward creating interfaces that are accessible, intuitive, and meet the needs of a diverse user base.

Cognitive dimensions framework

This framework provides a structured set of dimensions to evaluate the design's cognitive characteristics, such as learnability, visibility, and consistency, allowing us to create interfaces that align with users' mental processes so that we can enhance usability.

When making use of the cognitive dimensions framework, we approach it with empathy and a deep understanding of human cognition. We recognize that users have limited cognitive resources and aim to design interfaces that minimize cognitive load and support efficient mental processing.

The framework encompasses various dimensions, such as learnability, visibility, consistency, and others, each shedding light on a different aspect of the interface's cognitive impact. By systematically evaluating these dimensions, we gain insights into how the design influences users' perception, understanding, and interaction with the interface. Let's break those three down for deeper understanding:

- **Learnability**: Learnability refers to how easily users can grasp and understand the interface. It assesses how well the design supports users in acquiring new knowledge and skills to effectively use the system. High learnability means the interface has clear instructions, intuitive interactions, and interaction cues, and provides helpful guidance, allowing our users to quickly learn and become proficient in working with the application.

- **Visibility**: Visibility focuses on the clarity and prominence of the interface's elements and their affordances. It examines how well the design presents information and visual cues to users, ensuring that important elements are visible and easily recognizable. A high level of visibility means that relevant information, actions, and feedback are readily perceptible, allowing users to understand the state of the system and make informed decisions.

- **Consistency**: Consistency evaluates the uniformity and predictability of the interface's design elements and interactions. It examines how well the design adheres to established patterns, conventions, and brand standards, both within the interface itself and in relation to other familiar systems. Consistency ensures that users can anticipate how the system behaves, making their interactions more efficient and reducing cognitive load. A consistent interface reduces the need for users to relearn interactions and allows them to transfer their knowledge and skills across different parts of the system.

- **Findability**: Findability refers to the ease with which users can locate and access the information or features they are seeking within a system or interface. It focuses on ensuring that users can effectively navigate through the interface and locate desired content or functionalities without unnecessary effort or confusion. Findability plays a crucial role in enhancing the user experience by reducing frustration, saving time, and promoting successful interactions. When it comes to findability, heuristics can assist in evaluating the effectiveness of navigation structures, search functionalities, and information organization within the interface.

To effectively utilize the cognitive dimensions framework, we start by identifying the dimensions that are most relevant to the particular interface or problem at hand. We then assess each dimension by considering how well the design adheres to the principles associated with it.

During the evaluation, we observe and analyze how the interface supports users in acquiring new knowledge, how well it presents information and feedback, and how consistent it is in terms of interaction patterns and visual elements. We pay attention to factors such as the complexity of the interface, the ease of navigation, and the clarity of the information hierarchy.

By leveraging the cognitive dimensions framework, we gain a comprehensive understanding of the interface's cognitive strengths and weaknesses. This understanding allows us to make informed design decisions aimed at improving the interface's usability and user experience.

In summary, the cognitive dimensions framework offers a structured approach to assessing the cognitive characteristics of an interface. By applying this framework with empathy and expertise, we gain insights into how the design influences our users' cognitive processes. This understanding empowers us to create interfaces that are intuitive, efficient, and supportive of users' mental models, ultimately leading to improved usability and user satisfaction.

Wrapping up

The four evaluation models listed here, together with your new understanding of the Nielsen Norman Group's platform and model, should give you a solid grounding in the UX practice of heuristics. There will be times when you can only use one of the four, and there will be times when combining methods will be most valuable to your users and company. If you assess properly at the beginning, you'll be able to recommend and lead with confidence and provide first-rate reports on your findings when complete.

If your company has its own design system, these heuristic analytics can be a good indicator of your pattern library and its effectiveness – or not – in conveying the intended workflows and interactions required to successfully complete tasks.

Summary

Heuristics, as I think you can tell from this chapter, is a necessary toolset for any UX analytics practitioner or team to master and utilize. But it's not a standalone practice – it's part of the entire methodology that UX analytics encompasses.

You can do heuristics and then use data visualization to do your reporting. After you interview your stakeholder, you can start a heuristics review when told that the company wants to add new features to an existing product.

When your analytics via tools such as **Hotjar** show you where problems are arising with a group of users, you can test to verify that with a heuristics evaluation. The point is that UX analytics and the practice you and your team develop, fine-tune, and perfect is the sum of quite a few moving parts. It's a holistic process of best practices that involves people, data, and systems. It involves interaction and cooperation between teams that have differing agendas, but always remember – always – what the end goal is. And that's to ensure that the company is doing the absolute best it can to keep users and customers happy and engaged in working for and dealing with the company.

In conclusion and thanks

Well, we've made it! There was a lot to cover and I sincerely thank you for your time and attention. I hope you understand my decision to teach you UX analytics this way – as the broad practice that it is, and to ensure that you have a solid grounding in all the aspects, not just in the straight metrics you get from the tools.

If my years in this industry have taught me one thing, it's that everyone thinks differently and there's never one cookie-cutter solution to all the problems we face. When we understand UX analytics in this way, from this broad perspective, we're able to see where problems might be hiding and take care of them before they become insurmountable.

All my best wishes to you in your career, and if you'd like to connect on LinkedIn, please do so – connections and networks are key to long-lasting success.

Index

Packtpub.com

Subscribe to our online digital library for full access to over 7,000 books and videos, as well as industry leading tools to help you plan your personal development and advance your career. For more information, please visit our website.

Why subscribe?

- Spend less time learning and more time coding with practical eBooks and Videos from over 4,000 industry professionals

- Improve your learning with Skill Plans built especially for you

- Get a free eBook or video every month

- Fully searchable for easy access to vital information

- Copy and paste, print, and bookmark content

Did you know that Packt offers eBook versions of every book published, with PDF and ePub files available? You can upgrade to the eBook version at packtpub.com and as a print book customer, you are entitled to a discount on the eBook copy. Get in touch with us at customercare@packtpub.com for more details.

At www.packtpub.com, you can also read a collection of free technical articles, sign up for a range of free newsletters, and receive exclusive discounts and offers on Packt books and eBooks.

Other Books You May Enjoy

If you enjoyed this book, you may be interested in these other books by Packt:

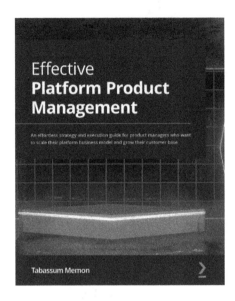

Effective Platform Product Management

Tabassum Memon

ISBN: 978-1-80181-135-4

- Understand the difference between the product and platform business model
- Build an end-to-end platform strategy from scratch
- Translate the platform strategy to a roadmap with a well-defined implementation plan
- Define the MVP for faster releases and test viability in the early stages
- Create an operating model and design an execution plan
- Measure the success or failure of the platform and make iterations after feedback\

API Analytics for Product Managers

Deepa Goyal

ISBN: 978-1-80324-765-6

- Build a long-term strategy for an API
- Explore the concepts of the API life cycle and API maturity
- Understand APIs from a product management perspective
- Create support models for your APIs that scale with the product
- Apply user research principles to APIs
- Explore the metrics of activation, retention, engagement, and churn
- Cluster metrics together to provide context
- Examine the consequences of gameable and vanity metrics

The Art of Crafting User Stories

Christopher Lee

ISBN: 9781837639496

- Leverage user personas in product development for prioritizing features and guiding design decisions

- Communicate with stakeholders to gather accurate information for writing user stories

- Avoid common mistakes by implementing best practices for user story development

- Estimate the time and resources required for each user story and incorporate estimates into the product plan

- Apply product frameworks and techniques for user story prioritization and requirement elicitation

- Benefit from the experiences, insights, and practices of experts in the field of user story mapping

Packt is searching for authors like you

If you're interested in becoming an author for Packt, please visit `authors.packtpub.com` and apply today. We have worked with thousands of developers and tech professionals, just like you, to help them share their insight with the global tech community. You can make a general application, apply for a specific hot topic that we are recruiting an author for, or submit your own idea.

Share Your Thoughts

Now you've finished *Unleashing the Power of UX Analytics*, we'd love to hear your thoughts! Scan the QR code below to go straight to the Amazon review page for this book and share your feedback or leave a review on the site that you purchased it from.

`https://packt.link/r/1804614742`

Your review is important to us and the tech community and will help us make sure we're delivering excellent quality content.

Download a free PDF copy of this book

Thanks for purchasing this book!

Do you like to read on the go but are unable to carry your print books everywhere?

Is your eBook purchase not compatible with the device of your choice?

Don't worry, now with every Packt book you get a DRM-free PDF version of that book at no cost.

Read anywhere, any place, on any device. Search, copy, and paste code from your favorite technical books directly into your application.

The perks don't stop there, you can get exclusive access to discounts, newsletters, and great free content in your inbox daily

Follow these simple steps to get the benefits:

1. Scan the QR code or visit the link below

https://packt.link/free-ebook/9781804614747

2. Submit your proof of purchase

3. That's it! We'll send your free PDF and other benefits to your email directly

www.ingramcontent.com/pod-product-compliance
Lightning Source LLC
Chambersburg PA
CBHW060526060326
40690CB00017B/3403